T0355944

TIME, MYTH AND MATTER

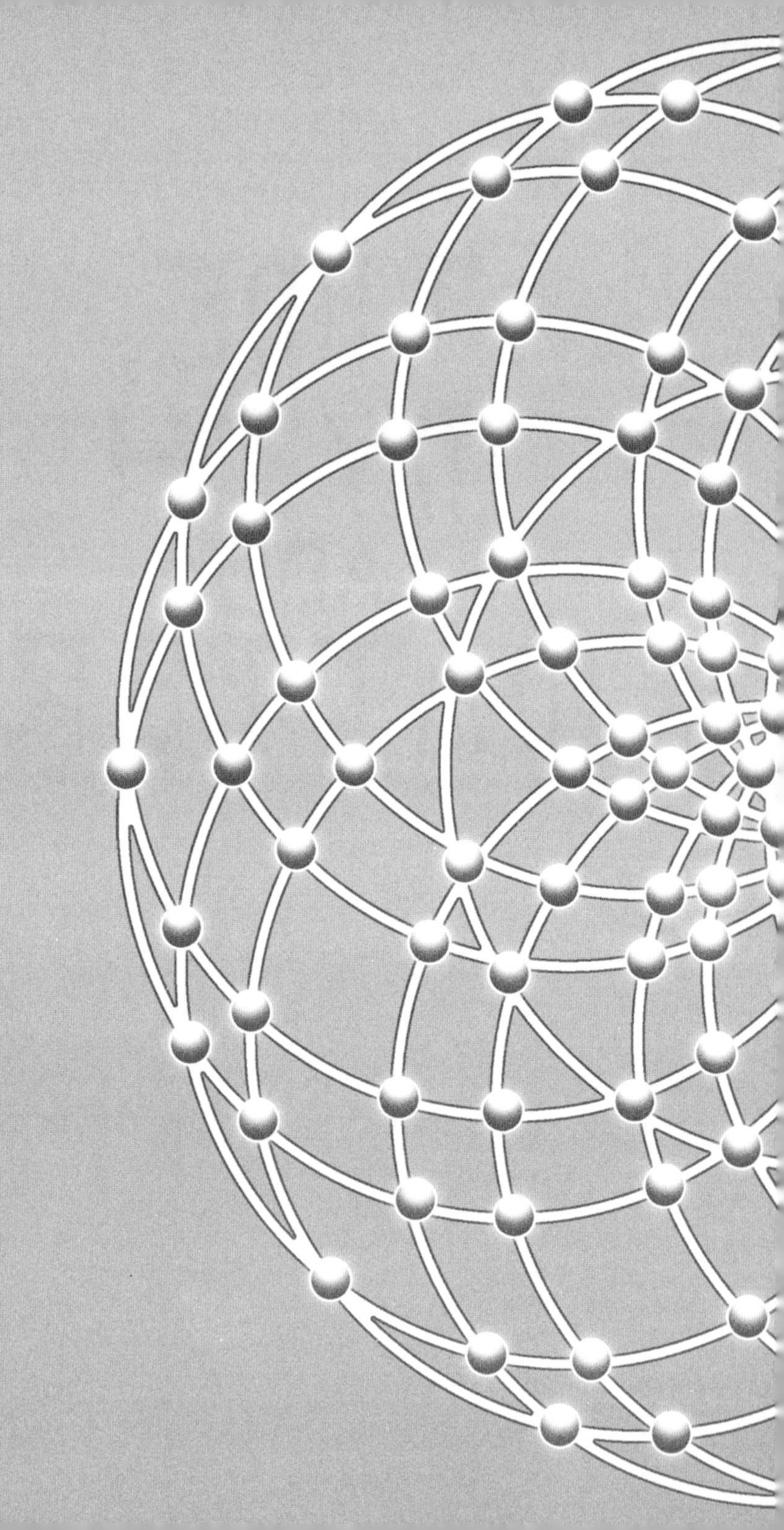

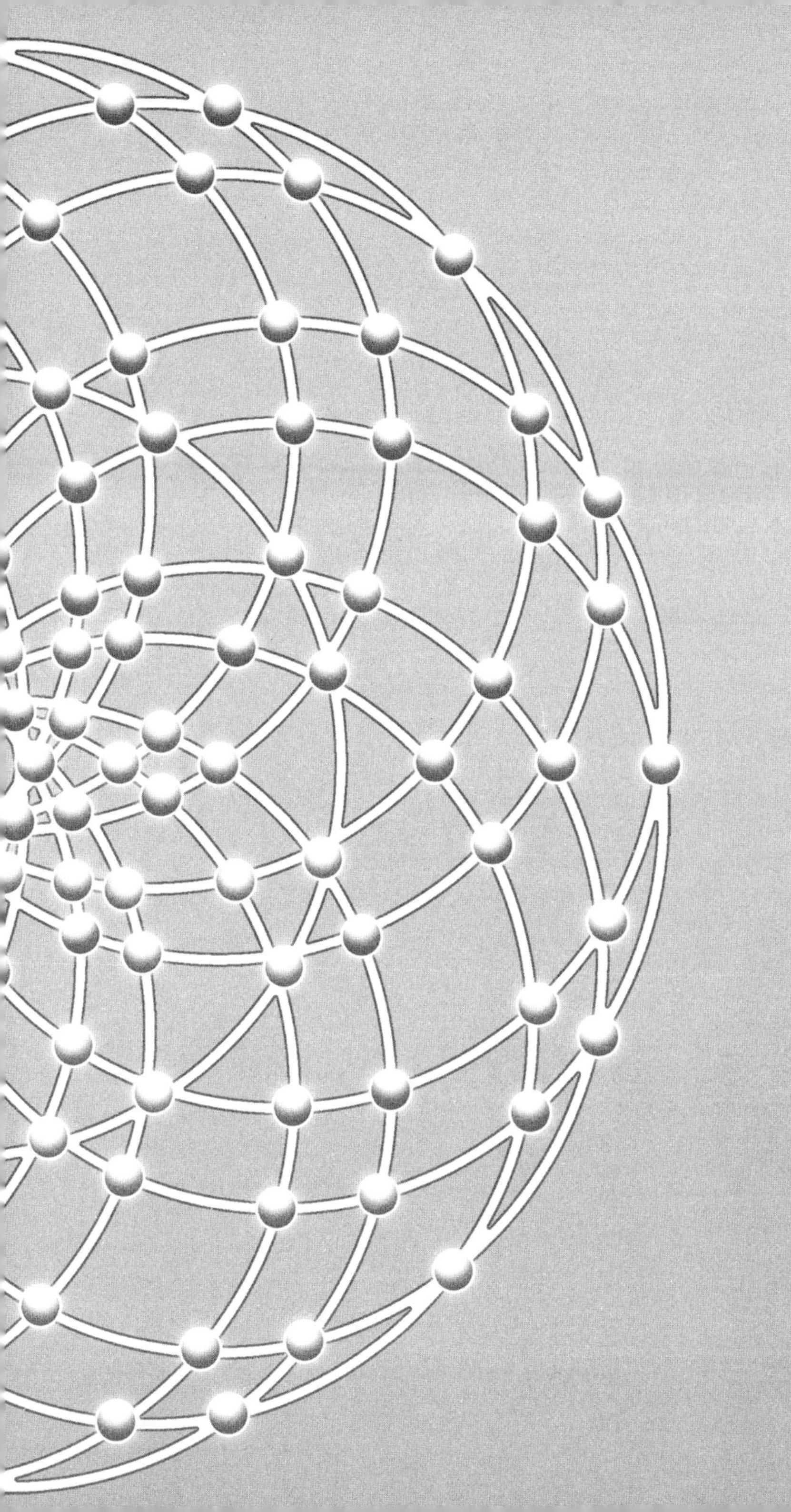

Time, Myth and Matter
LD Deutsch

Printed in China.

ISBN: 978-1-7361469-3-4

Published by Sacred Bones Books
Cover design by Sam Klickner and Brian Merriam
Layout by Mercy Correll, Sacred Bones Design

First Edition

1 2 3 4 5 6 7 8 9 10

All requests and correspondence can be addressed to:

Sacred Bones Books
144 N. 7th Street #413
Brooklyn, NY 11249

SBB-022

TIME, MYTH AND MATTER

ESSAYS ON THE NATURES AND NARRATIVES OF REALITY

LD DEUTSCH

WITH ILLUSTRATIONS BY SAM KLICKNER

SACRED BONES BOOKS

In loving memory of Mike Burakoff,
brilliant friend with whom I had the great joy
of discussing the many mysteries of reality.

It is probably true quite generally that in the history of human thinking the most fruitful developments frequently take place at those points where two different lines of thought meet. These lines may have their roots in quite different parts of human culture, in different times or different cultural environments or different religious traditions; hence if they actually meet, that is, if they are at least so much related to each other that a real interaction can take place, then one may hope that new and interesting developments may follow.

—Werner Heisenberg

CONTENTS

ACKNOWLEDGMENTS

I am incredibly grateful to everyone who has supported and encouraged the writing, thinking and living that went into this book. To everyone at Sacred Bones, thank you from the bottom of my heart for the years of friendship and support. To the extraordinary Carrie Schaff, without whom this book, nor the zines that preceded it, would exist—thank you is not enough. Your friendship is one of the great joys of my life. To my illustrator Sam Klickner, whose wonderful illustrations have brought so much life to these pages, thank you for your tremendous contribution to these essays. To my editor Rob Goyanes, thank you for your insightful and diligent work. To Elizabeth Youle, Jordan Williams and Brian Merriam, thank you for the riffing, insight into the material and design elements that have helped shape this book. To Victoria Bermudez, Emma Kohlmann, Chelsea Marks, Nina Hartmann, Marfisia Bel, M. Elizabeth Scott and my wider community of friends and peers, thank you for the years of conversations and explorations about the theories and ideas within these essays. To my family—my mother Cindy and my brother Jon, thank you for the countless ways you have helped support me throughout this process. And thank you especially to my father Andy, whose readership, insight, edits and advice have made an enormous positive impact on this book. To Bruce Parent, whose guidance through the realms of psyche has made all the difference in my life, thank you. Finally, I would like to thank the great, wild mystery that is reality.

A LETTER TO THE READER

I believe that, deep down, every person has a perspective on the ultimate nature of reality. Some are drawn to the question like a moth to a flame, while others are repelled by the unknowable that lies at the heart of the inquiry. But the question does not have to be frightening. It can be generative, joyful and vitalizing. If approached with playfulness and bravery, reality will play back with the inquirer, and a two-way channel between the material and nonmaterial may open. Consciousness plays a *real* part in the process of how reality arranges itself, and when the subject is met with respect and openness, a true, co-creative relationship may evolve.

These essays are attempts to participate in that endeavor. They were not written from the perspective of trying to *solve* anything, nor do they present any kind of unified theory. Instead, these essays come from, and exist within, a philosophy of instability. They are exercises in *reading* reality through various lenses and mediums. And although each is grounded in historical aspects of technoscience—in theories and "truths" that may change as the technoscientific process evolves throughout time—the phenomena illustrated within them is universal, timeless and may be applied in many different contexts.

This reading of reality may be considered a kind of hermeneutics, the art of interpretation. The religious studies professor Jeffrey Kripal writes that a major facet of hermeneutics is a "strange loopiness," a "paradoxical 'circle'" that transforms "both the read and the reader." Kripal emphasizes that in a true hermeneutical practice, there is no "stable 'subject' looking at or interpreting a stable 'object.'" Instead, "there is a single process that co-creates both the subject and the object *at the same time.*"[1] The work in this collection participates in that tradition.

In this effort, I have utilized theories and ideas introduced by the twentieth century analytical psychologist Carl Jung. To me, Jungian ideas provide an excellent, flexible basis from which to examine and engage with aspects of reality that lie both in the *outer world* and the *inner world.* Jung divides the material that makes up the inner world into three parts: the personal conscious ego, the personal unconscious and the collective unconscious. It is the relationship between the personal conscious ego (my own interpretive perspective) and the collective unconscious (the various episodes and patterns of history, technoscience and mythology) that is explored within these essays. One important Jungian theme that can be seen throughout the book is the idea that without a connection between the conscious and collective unconscious levels of psychological reality, we lose a vital natural link that is paramount to the health of our species. However, if I were to label it, I would actually call this book post-Jungian, rather than Jungian. That is because, although each essay has some Jungian concept as its foundation, the hermeneutical work that evolves from there builds upon Jung's ideas and develops them toward new aims. What results is an exercise in synthetic thinking, combining different existing ideas and perspectives in unique ways in order to generate new connections and possibilities.

The essays in this collection were first written between 2019 and 2024: "Pluto and the Mythic Dimension" in 2019, "Technomythology" in 2021, "The Myth of Matter: Part I" in 2022, "Myths + Models of Time & Timelessness" in 2023 (although the first iteration of this essay was written in 2015 and the second in 2018), and "The Myth of Matter: Part II" in 2024. All the essays that were written before 2024 were significantly revised and expanded in the process of compiling this book. It is my sincere wish that they be read as a warm invitation to develop your own joyful, generative, reality-reading process. Please remember that the ideas and theories within these pages can be understood by all. If at any point they ever feel a bit intimidating or overwhelming, please know that I also felt that way at one time. I knocked on the doors of these ideas for years before I felt like I truly understood them. I have tried, in these essays, to translate what I have seen and understood, and braid that together with what others have seen and understood, so that hopefully the doors may unlatch a little quicker for you, dear reader. Thank you in advance for your openness in reading what follows—I'll see you on the other side of the doorway.

PLUTO

AND THE MYTHIC DIMENSION

There is a mythic dimension of reality that permeates history throughout time. A narrative thread born of the collective psyche weaves itself into both mind and matter, revealing the two to be not only related, but equal expressions of one and the same thing. This dimension can be felt in the direct experience of consciousness, and takes shape in the stories that we tell each other and ourselves. Within its realm are certain historical moments that seem more deeply saturated with meaning than others, moments in which history and myth converge into one animate entity. These are the junctures that pivot the direction of humanity, changing something fundamental in the outer world as well as something intrinsic within us. The discovery of the planet Pluto was one such moment.

Pluto roared into existence in 1930, between the world wars and four months after the stock market crash of 1929. Its discovery was hailed as the scientific achievement of the century. The so-called ninth planet cracked open the universe, and its name selection brought with it the underworld, where the mythic Pluto reigned. Suddenly, the collective psyche was forced to shift, to make room not only for a larger solar system, but also for the re-emergence of the death god into the modern conscious awareness. These enormous changes foreshadowed an even deeper transformation of humanity's relationship to mass death and evil. The cultural synchronicities between the discovery of Pluto, with all the weight attached to its name, and the development of nuclear fission, leading to the eventual dropping of the atom bombs, would get recorded within the same breath of history. Pluto would prove an accurate augur for what was to follow and the world would be changed forever.

However, the further the story is pursued, the richer and more complicated it becomes. Pluto's significance extends far beyond the effect of its astronomical appearance and the events that followed. From its discovery in 1930 to its diminution to dwarf planet status in 2006, Pluto has mesmerized and challenged all areas of thought from the scientific to the symbolic. Drawing from its mythic origins, Pluto has come to be known as a metaphor for deep realms of the unconscious, which although overwhelming and frightening, yield great and meaningful treasure. The influence of this planet and its mythology endures to this day. Although the forms may be different, great changes that occur simultaneously within the collective unconscious and the outer material world fall into Pluto's lineage.

We are once again in a historical moment of tremendous transformation, with dire consequences looming. The way forward may lie somewhere in Pluto's story: darkness and wisdom leading back to one another.

The history of Pluto's discovery is bound tightly to the histories of Uranus and Neptune. In March of 1781, while searching the sky for double stars, the astronomer William Herschel came across a celestial object that he first thought to be a comet. He had actually stumbled upon Uranus, the first planet found with the aid of a telescope. The fact that an instrument of emerging technology was used in the planet's discovery is fitting, as Uranus was found during the peak of the Enlightenment. The accidental sighting would break open the boundaries of what was thought possible and rearrange our concept of outer space.

It did not initially occur to Herschel that what he saw could be a planet. At the time there was much public bias against the notion—up until then, all the known planets could be seen with the naked eye, and there was much cultural and religious significance assigned to a solar system that ended with Saturn. In fact, other astronomers had noted Uranus before Herschel, but it had always been classified as a star. Herschel was the first to notice that it was moving, unlike a star and not in the erratic way that comets tended to move. This object moved in an orderly path that was almost circular. It had to be a planet.

However, there were certain peculiarities in Uranus's apparent orbit. The path of the newly discovered body, as it orbited around the Sun, did not seem to adhere entirely to Newtonian laws of gravity. After exhausting other possibilities, the only explanation deemed plausible was that there was another, larger celestial object, somewhere beyond Uranus, whose gravity was pulling at Uranus and affecting its trajectory. This intimation sparked a great astronomical investigation, which eventually led to the discovery of Neptune by Johann Gottfried Galle in September 1846.

Although the gravity from Neptune did account for some of Uranus's abnormalities, it did not explain all of them. Furthermore,

Neptune appeared to have some orbital eccentricities of its own. And so there was hypothesized yet another planet, one even larger than and beyond the orbit of Neptune, whose gravity disturbed the paths of both Uranus and Neptune. This hypothesized planet soon became known as Planet X and thus the search for Pluto was born.

The case for Planet X was championed by Percival Lowell, a wealthy American who founded the famous Lowell Observatory in Flagstaff, Arizona. It's worth noting that the first part of Lowell's astronomical career was devoted to his fanatical belief that there were alien-made canals on Mars. Later, the images that so convinced him of this were shown to be optical illusions created by certain cratered patterns on the surface of the red planet. However, his lasting legacy would develop from his ardent search for the object that could cause such disturbances to the orbits of Uranus and Neptune.

Lowell searched for Planet X from 1906 until his death in 1916. When he died, an astronomer named Clyde W. Tombaugh took over the mission. Using Lowell's calculations as a guide, Tombaugh searched the proposed regions of space using a blink comparator, an astronomical tool that allows comparison between two different photographs of the same area of space taken at two different times. Then, around 4:00 p.m. on February 18, 1930, while examining the area surrounding Delta Geminorum (the eighth-brightest star in the constellation of Gemini), Tombaugh spotted what would become the last planet to be discovered in our solar system.

When it came to naming this new member of the planetary neighborhood, there were a few initial options proposed. Planet X was suggested as a permanent choice, as was the name Lowell. Janus, the double-faced god of new beginnings was also proposed because his two faces represented where humanity now stood in relation to the cosmos: one face looking back towards our Sun, and the other looking outward toward the vast unknown eternity of outer space. Minerva, Roman goddess of wisdom, was the strongest original contender, the option being seen as a way of adding a second feminine presence to the zodiacal cast.

Although Minerva ultimately did not make the cut, the naming process would not go without feminine influence. The honor eventually fell to an eleven-year-old girl from Oxford, England named Venetia Burney. Venetia had been studying Roman mythology in school and suggested to her well-connected grandfather that the new planet be named Pluto, to keep with the theme of the others. Her grandfather, a retired librarian from University of Oxford, telegraphed Lowell Observatory and proposed his granddaughter's recommendation. The choice of the name Pluto was announced on May 1, 1930, with the first two letters providing a lovely accidental homage to Percival Lowell. It is fitting that a young girl would give Pluto its name, as it is a young woman who features as the star of what is arguably the most plutonic myth of all—that of Persephone. Although Venetia Burney knew of Pluto's mythology, and that he was the god of death, there is no way she—or anyone else involved in choosing a name for the new planet—could have known how appropriate the suggestion would prove to be.

Pluto is the Latinized name for the Greek god Hades, lord of the underworld. Hades was the brother of Zeus and Poseidon; all were the sons of the Titans Kronos and Rhea. After the three younger gods overthrew their parents in the War of the Titans, they divided among themselves all things in existence. Zeus became the ruler of the skies, Poseidon of the seas, and Hades of the invisible underworld.

The word Hades translates as "the unseen one."[1] As a god he was associated with formlessness, invisibility and shadows, and was the only deity without a dedicated shrine. He was infrequently portrayed in art or pottery and his face is regularly obscured in the images of him that do exist. Often the depiction of his three-headed dog Cerberus is the only hint that the godlike figure in the image is in fact Hades. Because his depiction was such an uncommon occurrence, there is no lasting consensus about what Hades looked like. His personality also remains somewhat indefinable, and can really only be pieced together by observing the effect he has on the other mythological characters with whom he comes in contact. His reputation was one of sternness interspersed with the occasional act of altruism. Although he was unpitying, his appearance and function in a myth often involved establishing or maintaining some kind of necessary balance.

Hades' character and appearance remain so elusive in part because the ancient Greeks feared him greatly. They tried not to think of him too much, lest they might draw his attention. It is for this reason that we have so few artistic representations of him. The fear of his overwhelming power also prevented the ancient Greeks from using the name "Hades" directly except when absolutely necessary. Instead, they referred to him through euphemisms. Around the fifth century BCE, the god known as Hades started to be referred to as Pluto—the name change possibly taking such firm hold because of the fear surrounding the utterance of his original name. The Latin root of "Pluto" translates as "riches," "wealth" or "wealthy," and introduces an interesting new quality into this underworldly figure. Now a god who had primarily been identified with the fearful aspects of invisibility and death also becomes associated with nourishment and gift giving. This aligns well with the unique fact that Hades is the term for both the god and the realm over which he presides: the underworld. In other words, the *inner*world, here referencing inner Earth (that which is under the surface on which mortals dwell). As so, it is also the realm of all that the Earth provides, and later, as we shall see, also metaphoric of the inner world of man and the deepest layer of psyche. It is from this realm that all gifts emerge.

Despite all the nebulosity of Hades' character, there is one myth in which his desires are heavily featured. In it, we are provided some understanding of both his nature and the nature of the realm that bears his name, which are often inextricable. It is the myth of Persephone, of her descent into—and rise from—the underworld. Through her interactions with the god, and through the symbolism of the narrative at large, we are able to glimpse enough of Hades' shadow to intimate the sinewy laws of his domain.

The myth begins with Persephone and a party of other goddesses picking wildflowers in a field. Hades saw the group from afar and fell madly in love with Persephone. He approached Persephone's father, who happened to be his brother Zeus, and asked him for her hand in marriage. Wishing not to offend his brother but knowing well that Persephone's mother Demeter, goddess of Earth's fertility, would never approve of the marriage, Zeus neither gave nor denied his consent. Hades interpreted this gesture as Zeus 'not saying no' and so, emboldened by his brother's implied compliance, set out to abduct the virginal goddess.

While in the field, Persephone came upon a narcissus flower with a hundred heads. Some say the flower was placed there as a favor to Zeus by Gaia, an Earth-goddess ancestress of Demeter. The flower, which was unlike any Persephone had ever seen, mesmerized the young maiden and she plucked it. Suddenly, from the place where the flower had stood in the Earth, a great chasm ripped open and there appeared a portal to the underworld. From it, Hades emerged in his chariot and dragged a screaming Persephone down into his realm to be his queen.

Back on Earth, Demeter couldn't find Persephone. When the Earth goddess realized that her daughter was missing, she became consumed with grief and began to search the world over for her. As Demeter was the goddess of fertility, all living things on the planet ceased to grow because of her despair. Earth entered a perpetual winter, and became barren and unyielding. After enduring a challenging saga of her own, Demeter came to learn of Persephone's location from Helios, the all-seeing sun god, who admitted that he had witnessed Hades abduct her. Demeter was furious, but instead of going to Olympus to confront Zeus directly (whom she had discovered was also involved), she continued to roam the Earth despondently until all life teetered on the verge of extinction.

Fearing for all life on Earth, Zeus tried begging Demeter to return to Olympus but she refused. So, resigning himself to the only remaining option, Zeus sent his son Hermes down into the underworld in order to retrieve Persephone and return her to her mother—on the one condition that she had not yet tasted the food of the dead. If Persephone had tasted such food, she would have to remain below. Demeter was overjoyed by the decision; Hades was dismayed. Just then, a gardener from one of Hades' fruit fields announced that he had seen Persephone eat six seeds of a pomegranate from a tree in the orchard. Persephone was still taken to meet her mother but upon reaching the surface, the truth of the pomegranate was revealed. When Demeter heard of this, she became hysterical again. Knowing that all living things would perish unless Demeter was reunited with her daughter, Zeus came to resolve the matter. A compromise was reached: Persephone would live two-thirds of the year on the surface of the Earth in the company of her mother, and one-third of the year in Hades with her husband, where she would rule as queen of the underworld. The Earth would have eight months of beauty, growth, spring and harvest—and four months of degeneration, death and winter.

Psychologically, Persephone's tale is often interpreted as one of *initiation*, of having naivete ripped away and venturing through dark, unconscious and often frightening aspects of the transpersonal psyche—of becoming involved in something inherently overwhelming, but something that eventually leads back to wholeness. Persephone falls through an unknown crack in the known system, into a realm she could have never imagined. However, she soon becomes queen of the very realm into which she was reluctantly dragged. In fact, once queen she acquires the reputation of being merciful and gracious. It is only through her knowledge of the underworld, and the integration of that knowledge into her larger Self, that she gains her own personal power and the power of mercy. In Greek mythology, she is the only one allowed to live in both the upper and lower worlds, the upper and lower dimensions of reality. Thus, she becomes the ultimate image of a unified whole—although the unification was acquired through suffering.

In terms of understanding the symbolism of Pluto, one of the most important parts of this myth may be Persephone's descent. Through the act of picking a flower that was placed there for her to pick, Persephone endures an involuntary trauma. But the preexistence of the flower also implies that she is somehow destined to lose her innocence this way. She is taken to the underworld by a god whose presence is inherent in the flower itself, but is hidden from her viewpoint (another example of Hades' invisibility). This is an aspect of Nature at work. She is unconscious of the depths and unaware of how close she stands to them. Yet this is her fate. She must join in union with Hades. She must become profoundly acquainted with the regions of reality that are darkest, most mysterious and hardest to integrate. She cannot remain ignorant. Through this myth the depths announce that they are a part of Nature, and that they will be known.

Pluto's appearance in the myth of Persephone was as forceful as the appearance of his namesake planet into the historical, material realm, and each required a complete transformation of consciousness. The celestial object and mythological god both reigned in the realms of utter darkness—one at nearly four billion miles from the Sun, and the other in the underworld. When the decision to name Pluto was made, it could not have yet been known that mankind itself would soon go through its own plutonic initiation. The world was between the First and Second World Wars, and humanity was passing through a portal in which what was deemed possible was changing drastically. A chasm was opening: a hidden presence much like the one lying in wait within the narcissus was surrounding history at that moment, threatening the very material of reality itself.

Pluto was discovered and named in 1930. The neutron was discovered by physicists in 1932. Also in 1932, experiments proved that by bombarding atoms with accelerated protons, a nuclear transformation could be induced. In 1935, it was discovered that this process produced a greater variety of results when neutrons were used in the place of protons. Then, in December 1938, while experimenting with bombarding an assortment of different elements with neutrons, the scientists Otto Hahn and Fritz Strassmann discovered what would become nuclear fission and unleashed atomic energy for the first time.

While most of the elements tested were transformed only a little by being bombarded with neutrons, it was discovered that uranium nuclei changed enormously, dividing into two approximately equal parts. Found among the uranium's debris was radioactive barium, which had an atomic number of 56. The barium weighed less than the original uranium nucleus, which has an atomic number of 92. This became the first experiment to confirm Einstein's theory of equivalence between mass and energy ($e = mc^2$). The process was named "fission" after the biological term for cell division. Further experiments showed that not only did this fission cause an initial release of enormous amounts of energy, but also caused the release of other neutrons, which could possibly collide with other atoms and potentially cause a self-sustaining chain reaction of ever-increasing amounts of energy.

In 1941, while at the University of California, Berkeley, the chemist Glenn T. Seaborg and a team of scientists experimented with bombarding uranium in a cyclotron, a particle accelerator that uses an alternating electric field which is shaped into a spiral by a static magnetic field. When uranium was bombarded in such a way, it produced neptunium, an element with an atomic number of 93. Although neptunium was initially discovered in 1939, the experiments done at Berkeley showed that in the cyclotron, the neptunium subsequently began to decay, revealing an entirely new element. With ninety-four protons in its nucleus, this element was heavier than any that had been encountered before.

When deciding what to name this newly found element, Seaborg and his team followed history's example. The planet Uranus was discovered at the end of the eighteenth century, Neptune was discovered in the nineteenth century and Pluto was discovered in the twentieth. Uranium was found in 1781, eight years after the discovery of Uranus. Choosing to name uranium after Uranus was an easy decision at the time as both discoveries epitomized the spirit that defined the Enlightenment. No new elements were found until the atomic experiments that started in the 1930s yielded neptunium in 1939. When it came to naming neptunium, science looked back to the last time an element was named and followed suit. So, when it came to the heavy thing found in December 1940, Seaborg almost had no choice but to call it plutonium. It is a strange synchronicity that both uranium and plutonium were found so close to the discovery of their namesake planets, and that it was these two elements that would be used in the atomic bombs—it was as if the elements and planets were two ends of the same pole.[i] However, uranium occurs naturally, whereas plutonium only exists because of the interference of man. Reactor-grade plutonium will spontaneously ignite when exposed to air. Because of this, plutonium must be surrounded by nitrogen at all times. It is so dangerous that its toxic dose is measured in millionths of a gram.

Plutonium was the active ingredient in the atomic bomb dropped on Nagasaki on August 9, 1945. Uranium was the active

---

i The god Pluto's association with that which is hidden relates quite interestingly to the history of plutonium. The efforts and experiments done in the creation of the element were performed in such secrecy that the research that went into it was never recorded in print. The findings were not even published until after the bomb was dropped on Nagasaki. The actual dates and times of discovery are only known thanks to a correspondence of Seaborg's.

ingredient in the bomb dropped on Hiroshima three days prior. The death toll of the combined bombings would exceed 226,000 people. About half of the deaths in each city happened on the first day. The rest were from acute effects of radiation.

Through the nuclear experiments done in the 1930s, mankind met an energy force that had never before made it to the earthly material plane. It was as though by harnessing atomic energy, humanity had spontaneously encountered, or co-created, the collective mythic counterpart of such energy. If such a power had made it to the realm of matter, the myth needed to materialize as well. We met a new god in the form gods take in the modern world: a planet.

After this, there could be no claims of absolute knowledge anymore. The very unit of matter was broken apart. Modernity had rendered reality obsolete, leaving a boundarylessness, a formlessness, a limitless repeating chain of reaction. If the atom, the unit of life, falls under Zeus's domain, then the splitting of it would invite Hades to the table.

Up until the late 1970s, Pluto held fast to its position as the ninth planet, even though successive astronomical observations yielded increasingly smaller estimates of its size (throughout its lifetime as a planet, Pluto would go from being hypothetically bigger than Jupiter, to in reality being smaller than Earth's moon). Although a few prophetically minded astronomical thinkers had already begun to predict Pluto's fate, it wasn't until the *Voyager* missions of 1977 that things started to really shift. As the *Voyagers* started to send back detailed images of Jupiter, Saturn, Uranus, Neptune and the celestial objects that surround them, we began to see a more complicated and diverse picture of our solar system. The new information from these images sparked a deeper interest in properly defining what exactly made up our planetary neighborhood. Suddenly, Pluto's description was fitting in better with groups of asteroids and comets rather than other planets.

In 1978, Pluto's moon Charon was found. Named after the psychopomp who ferried souls across the river Styx, Charon the moon turned out to be nearly the size of Pluto itself. The two were discovered to be in a tidal lock, meaning that they always show the same face to each other. Already feeling the changing definitional tides, many Pluto-as-planet enthusiasts cited Charon as categorical evidence of Pluto's planethood, arguing that only planets have moons. Unfortunately, when the asteroid Ida was discovered with its moon Dactyl in 1994, this argument had to be given up.

By the 1980s our understanding of our solar system was changing. As technology progressed and astronomical instruments became more refined, we began to reclassify the existing planets even further. Mars, Earth, Venus and Mercury became known as terrestrial planets because they are dense, rocky and small. Neptune, Uranus, Saturn and Jupiter became classified as Jovian planets, because they are gaseous, large and ringed. As the other planets began to reveal their specific natures, it only served to highlight how different Pluto truly was.

Then in 1992, astronomers Jane Luu and David Jewitt discovered the first object of the Kuiper Belt that was neither Pluto nor Charon. The Kuiper Belt is the name given to the vast icy region of the solar system beyond Neptune, which is home to Pluto, and would reveal itself to be host to many other icy celestial bodies as well. Although this first object, known as 1992 QB1, was not similar enough to Pluto to threaten its definition, the astronomers soon found objects that did share some of Pluto's traits, such as an elongated orbit. Luu, Jewitt and others would go on to find other celestial objects of similar size and mass to Pluto. Then, in 2005 a team of astronomers from Caltech led by Mike Brown discovered Eris, a trans-Neptunian celestial body with an estimated mass larger than Pluto's. From then on, all bets were off.

In January 2006, NASA's *New Horizons* launched its mission to image Pluto and its moons, as well as other members of the Kuiper Belt. At the time, the grand-piano-sized spacecraft was the fastest man-made object to ever leave Earth. It passed through the Jupiter system in February and March of 2007, which provided it with a gravity boost, allowing it to fly by Pluto in 2015. On board the craft was an ounce of Clyde Tombaugh's ashes and a "dust counter" named after Venetia Burney—the astronomer who first spotted Planet X and the young girl who named it Pluto, respectively.

Also in 2006, the International Astronomical Union met in Prague to discuss, amongst other topics, what to do with Pluto's planetary classification. At first, the IAU considered expanding the solar system to twelve planets that would include Pluto and Charon as twin planets. That proposal was rejected. Instead, the IAU decided to redefine a planet to mean a spherical celestial object that orbits the sun, and clears the environment around its orbit by the sheer force of its own gravity. It was this last classification criteria that sealed Pluto's fate. Because Pluto does not have sufficient mass to influence the other Kuiper Belt objects, it would be henceforth known as a dwarf planet, one of a flock of other similar bodies.

The scientific and popular world erupted in controversy over the demotion. Opponents pointed out that only one-tenth of the members of the IAU were actually present for the deciding vote. The timing was also conspicuous. Debates about Pluto's planethood had long been underway, but the IAU waited until after *New Horizons* launched to make their formal decision. Many felt that this was an in-community slap in the face to Plutophiles. But whatever the reason, Pluto went from being the smallest of the planets to one of the largest of the Kuiper Belt objects.

There are many symbolic ways to read Pluto's turbulent history. The guardian who stood between our solar system and the rest of the universe had been removed from its post. If we follow the connection between the planets and the gods with whom they share names, the psychological reality becomes apparent. How are we to think of the nonphysical energies that move us? That orbit around us? How do we think about the forces that affect our lives in large but invisible ways? If a planet gets demoted, what happens to our collective relationship to that planet's corresponding mytheme?

The psychologist James Hillman wrote that archetypal material, which is constellated mythic material originating from the deepest layer of the collective unconscious, participates in determining its own expressive terms as an aspect of its own *self*-definition. Hillman, a poetic thinker who founded the psychological movement known as Archetypal Psychology, was a Jungian analyst whose work focused on the ways in which mythic archetypes organize the innermost stratum of psychic life. He explains that the metaphors chosen for describing archetypal concepts and systems are in fact intrinsic parts of those concepts and systems themselves. In Hillman's view, the human ego is relativized by the *archai*, which are fundamental transpersonal aspects of psyche that pattern and determine how humans construct meaning out of experience.[2] Hillman argues that any kind of archetypal material that makes its way into collective conscious awareness participates agentively in defining itself. Taking this into consideration, naming something becomes a *real* act and not just the application of a label; the name brings something into reality. In the case of Pluto, it is a reality of unimaginable depth, initiation and resurrection.

Hillman writes "there is no time in the underworld,"[3] and just as there is no time in the underworld, there is no time in the unconscious. And as the underworld is inaccessible to those who live on the surface, so too is the unconscious inaccessible to consciousness. However, there are some instances of activity that originate from the depths of the unconscious and burst forth into conscious dimensions of reality. We can feel the existence of the unconscious much like an orbital effect, similar to the originally proposed influence of Pluto on the orbits of Neptune and Uranus. Although we cannot know the unconscious directly, we become aware of its actuality due to its influence on other more conscious aspects of psyche. And just as Pluto proved to be a cluster of entities instead of one large one[ii], our relationship to the unconscious shifts from form to fleeting form, from coagulation to dispersion through the continuous act of relativization. This analogy may be taken one step further into the personal if the Sun may be considered a proper symbol for the ego. In the same way that our understanding of Pluto went from being one large body to a collection of smaller ones, so too do we intimate the existence of a multitude of larger unconscious forces (however slippery or shape-shifting they may be) by observing their effects on the personal aspects of our consciousness.

We are also reminded here that processes within the unconscious may also end up being quite different than what the conscious psyche perceives them to be. The original hypothesis that

ii The original prediction was that the mass of Planet X was equal to the mass of seven Earths.

fueled Percival Lowell's search was that there was a planet beyond Uranus and Neptune that was large enough to affect the orbits of both. This turned out to be false. More sophisticated instruments aboard *Voyager 2* showed that Lowell had overestimated the mass of Neptune, and that when the correct mass was used, there was no irregularity in the planet's orbit.

And like this proposed planetary influence, so too is the mythological Hades seen most clearly through his effect on other Greek mythological characters. The dragging down into depth from more terrestrial aspects of consciousness is inescapably destructive to the ego. The ego cannot understand why the trauma is occurring at the time of the event—we see an embodiment of this in Persephone. However, Hades is also known as a bringer of balance. And just as in the myth of Persephone, something in the myth and history of the planet Pluto is hinting that humanity is being initiated, but at the end of the initiation there is the possibility for wholeness. The unconscious may be seen as an obscure cluster of itself, a thing holding together. If Pluto can stand for a moment as the modern shadowy ambassador of the unconscious, then the coming into consciousness, and then the distortion and recession detailed in the planet's taxonomical history, aligns very well with the nature of the unconscious itself.

The twentieth century brought the planet Pluto into existence, and the twenty-first took it way. Upon Pluto's arrival, the world became initiated into a reality that could not have been imagined generations prior. Humanity began to see extraordinary advancements in technology, unimaginable death and the mysteries of matter revealed. The plutonic aspect of consciousness burrowed its way out into the light of day and with it came irreconcilable realities.

At the time of discovery, the public's digestion of the event had plutonic undertones as well. When Gustav Holst wrote his suite *The Planets* in 1916, inspired by the planets and the gods they were named after, the solar system then stopped at Neptune. After the discovery of Pluto, Holst set to write a movement dedicated to it. Fittingly, a stroke prevented him from ever finishing it. The writer

J.F. Martel also points out that Disney, the monolith of capitalist mythmaking, named their new dog character Pluto in 1930, right after the new planet received its name. Martel suggests that this move may have been an unconscious attempt to distract the world from understanding what exactly was being brought into being through the discovery and naming of the planet Pluto.[4] Instead of corporate America's catlike reflexes taking advantage of the times and zeitgeist, it may have actually been responding to something much deeper.

It has been said that our past decade (the early 2010s through the early 2020s), shares striking similarities with the 1930s: both eras include a rise of public racism and xenophobia, fascism, stock market volatility and banking crises, increased homelessness, rising global trade tensions, and an ever-steady increase of technology permeating society and upending entire industries. When I wrote the first version of this essay in 2019, I speculated if we too would have our own augural outer space discovery. At the time, I wrote this:

> I wondered as I began this essay, if we would have our own plutonic moment, our own new moment of dark discovery. If so, what would it portend this time? I believe we have been answered with the first imaging of the black hole, a near-incomprehensible picture of nothingness. Might this incredible astronomical feat have something to tell us about what's happening deep in the collective unconscious? Might there be some aspect emerging that looks like a black hole?

Less than a year later the COVID pandemic hit, but perhaps more pertinent is that a few short years after that the world saw an explosion in artificial intelligence, which scientists and philosophers now view as an existential threat to humanity unparallelled since the development of the atom bomb. In fact, the comparisons between the current quest to develop artificial general intelligence, or AGI, and the development of nuclear weapons during the 1930s and 1940s are numerous and often cited. During his keynote speech at the opening of the Stanford Institute for Human-Centered Artificial Intelligence in 2019, Bill Gates likened artificial intelligence to "nuclear weapons and nuclear energy," saying that they are "both promising and dangerous."[5] The computer scientist and cognitive psychologist Geoffrey Hinton, who is considered one of the "godfathers" of AI and who, along with two colleagues, won a Turing Award in 2018 for engineering and philosophical breakthroughs in neural networks,[iii] left his job at Google in May 2023 in order to speak more freely about the dangers of the technology he helped to invent. Since then, he has been explicit about his fears. Hinton warns that advanced AI is on its way to becoming smarter than humans, and if it were to get *much* smarter than us, it could (and would) figure out a way to kill us. As he puts it, "there are very few examples of a much more intelligent thing being controlled by a less intelligent thing."

Hinton, and others, make a marked distinction between this kind of fear, which is termed "existential" because it deals with the domination or total eradication of the human species by superintelligent AI, and the fear that AI technology could end up in the hands of wrong actors as a means of carrying out evil acts (something the world is already seeing). This existential threat usually applies to the potential development of AGI, which is any AI that can think, understand, adapt and create at or above the human cognitive level. In a CNN interview in 2023, Hinton made the direct comparison between AGI and nuclear weapons, calling for a global alliance in terms of regulation and safety. "It's like nuclear weapons," warned Hinton, "if there is a nuclear war, we all lose, and it's the same if these things take over."[6]

Hinton has come to regret his contribution to the field, but consoles himself with what he calls the "normal excuse: If I hadn't done it, somebody else would have."[7] There are, of course, also stories of scientists who contributed to the atom bomb coming to regret their involvement. Albert Einstein felt remorse for writing a 1939 letter to President Roosevelt suggesting that the (then) recent scientific breakthroughs regarding uranium and nuclear chain reactions could conceivably lead to the construction of "extremely powerful bombs of a new type."[8] The letter is considered by many to be the impetus for the American atomic bomb project. After the bombs were dropped on Nagasaki and Hiroshima, J. Robert

iii Neural networks are machine learning models modeled on the neuronal activity in the human brain, which serve as the computing architecture for advanced AI.

Oppenheimer told President Truman that he felt as though he had "blood on his hands,"[9] and less than ten days after the bombing of Nagasaki, Oppenheimer wrote a letter to the Secretary of War Henry Stimson calling for the ban of nuclear weapons. Both Einstein and Oppenheimer admitted that their involvement with the bomb was spurred by the fear that the Germans would develop one first, an arms-race sentiment echoed by Hinton—only this time instead of an atomic bomb it is AGI, and instead of Germany the adversary feared is China.

Hinton isn't the only one sounding the alarm. After OpenAI revealed a new powerful model of ChatGPT on March 14, 2023, a growing chorus of concerned voices began calling for action to mitigate the growing existential risk posed by AI. A slew of essays and open letters signed by leading scientists, professors and tech leaders began popping up, each with a different phrasing of similar worried sentiments. On March 22, 2023, the Future of Life Institute released "Pause Giant AI Experiments: An Open Letter," calling for a six month pause on training any AI systems that were more powerful than the ChatGPT-4 model that had just been released. The letter, which was signed by Elon Musk, Steve Wozniak, Max Tegmark and many others (although not by many people directly involved with major AI organizations), prefaces its suggestion of a pause with the questions "*Should* we develop nonhuman minds that might eventually outnumber, outsmart, obsolete and replace us?" and "*Should* we risk loss of control of our civilization?"[10]

On March 29, 2023, *Time* magazine published an essay by artificial intelligence researcher and decision theorist Eliezer Yudkowsky criticizing the Future of Life Institute's open letter, explaining that he refused to sign it because he thought the letter was "understating the seriousness of the situation and asking for too little to solve it." Yudkowsky calls for the shutting down of all advanced AI projects, saying that many researchers like himself believe that if an AI system smarter than humans were to be built "under anything remotely like the current circumstances ... literally everyone on Earth will die." Yudkowsky warns that the "thresholds" AI would need to cross in order to be considered smarter than a human might not even be noticeable, and that we may very well be "cross[ing] critical lines without noticing."[11]

Another letter, released by the Center for AI Safety on May 30, 2023, had a much sparser statement. The letter consists of one sentence that reads:

> Mitigating the risk of extinction from AI should be a global priority alongside other societal-scale risks such as pandemics and nuclear war.[12]

This brief letter was signed by the CEOs of three of the top AI organizations, Demis Hassabis of Google DeepMind, Dario Amodei of Anthropic and Sam Altman of OpenAI. Altman, who became CEO of OpenAI in 2019 and has become somewhat of a poster-person for the AI industry, is a very interesting character in this discussion. During a conversation with *The New York Times*, also in 2019, Altman casually compared OpenAI's quest to create AGI to the Manhattan Project, saying that the two shared "a level of ambition." During the same conversation Altman paraphrased Oppenheimer when he said, "Technology happens because it is possible,"[13] implying that he felt the same way about the inevitability of AGI as Oppenheimer did about the creation of the bomb. Altman also said in the interview that although he believed that AGI would produce global prosperity, he was also worried that it could destroy the world. Oppenheimer also once considered the possibility that the atomic bomb could destroy the world—if the extremely high temperatures released by the bomb were to explode the hydrogen in the ocean and in the air (which would lead beyond fission into nuclear *fusion*, unleashing an atomic explosion in the atmosphere). Though calculations ultimately led Oppenheimer and his team to be satisfied that the chances were low enough that this would not happen when the atomic bomb was detonated, the chances were never zero.[14]

Cade Metz, the interviewer from *The New York Times* who spoke to Altman in 2019, spoke to him again in 2023. In his 2023 piece, Metz wrote that in 2019, Altman's words sounded like science fiction, but reflecting back on his words, Metz wondered if they weren't actually prophetic.

There was one other detail that Altman mentioned to Metz: that he, Altman, and Oppenheimer share a birthday.

In a May 2023 blog post on OpenAI.com, Altman and two other OpenAI executives once again compared their technology to nuclear weapons, calling for an international institution, along the lines of the International Atomic Energy Agency, to monitor the safety of future AI development.[15] The comparisons between nuclear weapons and AI became so numerous that in June 2023, the *New York Times* ran a twelve-question quiz that asked the reader to discern which quote was about which.[16] The interchangeability between the quotes is remarkable. When I took the quiz while revising this essay, I got five out of twelve correct.

A black hole is an area of spacetime where a tremendous amount of matter compacts into a very small space. The matter is so dense and packed that its own gravity punches a "hole" in the fabric of spacetime, so that nothing that falls into it, not even light, can ever escape. The center of a black hole, called a singularity, is an area of infinite density, into which all the compressed matter collapses. Around the singularity is the event horizon, a sphere-like, virtual boundary that acts as the point of no return. Were an entity to find itself near the event horizon of a black hole, it may find itself *crossing critical lines without noticing*, and once energy or matter has crossed this threshold, it is gone. No one knows what happens to something when it slips into the grip of a black hole. It is the very end of spacetime as we know it.

A black hole cannot be seen, but because all light that touches a black hole disappears into it, there remains, just outside of the black hole's boundary, an extraordinary amount of light. It is this light that was captured in the image of the black hole Messier 87, located in the Virgo constellation, in 2019. The light, seen bending through the phenomenon of gravitational lensing as it is sucked into the blackhole, was imaged by the aptly named Event Horizon Telescope, through a process called Very Long Baseline Interferometry, which electronically connects telescopes all over the world to create one massive telescope system. Since 2019, the Event Horizon Telescope has also imaged the supermassive black hole that sits at the center of our own galaxy.

A black hole is far more intangible than a planet, even one whose definition changed and devolved over time like Pluto's. In the same way, the threat to humanity from AGI is more intangible (at this point) than the threat of nuclear war. And yet, both sets require a certain thinking of the unthinkable, and a definite acceptance of new and strange terms to reality. Neither nuclear weapons nor AI can be de-invented, and both hold similar powers to destroy the world—powers that would certainly compound were we ever to allow AI to control the use of nuclear weapons. And yet in both, we see the same emphasis on invisibility, the same drive toward progress without corresponding caution, and the same desire to master nature through technology.

In a 1959 article that appeared in *The American Weekly*, the Nobel Prize winning writer Pearl Buck spoke with another Nobel Prize winner Arthur Compton, the former director of the Manhattan Project, who was responsible for hiring Oppenheimer to head up the task of building the bomb. The opening paragraph of the article is as haunting as it is striking:

> The most important thing that faces us all today is the atom. On the one hand, it threatens us with annihilation; on the other it holds out the promise of a life of plenty, surpassing our wildest dreams. How did we get into this desperate situation? Is it within our power to choose between the alternatives? If so, how do we proceed and what do we do?[17]

If you were to replace the words "the atom" in the first sentence with the words, "artificial intelligence," this exact same paragraph could have been written today, in 2024.

Where do we go from here? What can we learn from the great changes humanity has already undergone? The reality that Pluto ushered in revealed an indefiniteness in our material nature. Could the reality being ushered in by the black hole reveal a definiteness in our nonmaterial nature? Both realities may have to be accepted, even though they may seem impossible to reconcile with the collective mind frame we have inherited. We may never be able to see a stable reflection of the truth *because the truthful reflection is not stable.* The bottom has no bottom. We must learn to find balance upon groundlessness. We must allow humility to lead to transcendence, and allow transcendence to lead back to humility. We have to accept reality and let it change us.

Persephone may be our greatest teacher in these present times. She pulls up the flower and with it, all of consciousness. She learns the truth of the underworld, and through the acceptance of its realm she becomes unified. She gains the capacity to move through dimensions of reality that she once found terrifying and inconceivable. The laws which govern her existence change drastically and permanently, all in one instant. And although there is a painful transition period to her new reality, she *does* make the transition. Something else has entered and forever altered the picture: Persephone has penetrated the depths just as the depths have penetrated her. She swallows the pomegranate seeds and metabolizes into herself something of the underworld. Yet no doubt a morsel of herself was to be found there in the first place, because she becomes a *queen*, a concentrated image of sovereignty and care. She is no longer only a daughter of Demeter; she is now a merciful ruler and her mercy is evidence of both growth and power.

Once Persephone performs the radical acceptance that is asked of her, she finds that she is able to integrate her new understanding. She no longer belongs entirely to the world she was born into, nor to the invisible world with which it is braided (as Hillman wrote "Hades is not an absence, but a hidden presence—even an invisible fullness"[18]). Yet she is at home in both, and more than that, recognizes the one within the other. This is uncharted territory; an entirely new way of moving through reality. Her experience is unprecedented, and the ecosystem in which she now lives includes an invisible, fear-producing dimension that holds within it fruits and riches. These hidden, potentially horrifying aspects of reality become naturalized through Persephone. She transcends her previous worldview and no longer privileges the upper-world over the underworld because she knows that, at each and every moment, the presence of one pervades and permeates the other. Persephone thus becomes a bridge, a living example of sacrifice and grace.

Persephone lets go of the life she once knew, the one governed solely by a mother principle replete with fertility and promise. She is abducted unwillingly into a future that she would certainly have once preferred not to have. However, it is this very abduction that allows Persephone to transcend the identity of daughterhood into one of self-determination and adaptation. In her beautiful book *The Kore Goddess*, Saffron Rossi reminds us that Persephone is the example par excellence of the Kore; those virginal maiden goddesses that appear throughout various mythological traditions. Before she is abducted, Persephone is often referred to simply as Kore or the Kore, becoming namable as Persephone only when she splits from pure identification with her mother Demeter.

The Greek root for the word Kore translates into "vital force," as that which produces life.[19] Rossi expands on this, explaining that although the Kore is most often imaged as a young maiden, the word itself "connotes youthfulness, not a particular number of years."[20] However, this vital force is different from the limitless fecundity that is seen in the Great Mother, which Rossi contends has been the prevailing feminine archetype in the West. Rossi writes that the indiscriminate abundance of the Great Mother "stands in contrast to the focus and choice of the Kore." Unlike the Great Mother, the "Kore makes sacrifice, differentiates and incubates the unique."[21] The main feature of the Kore, as Rossi explains it, is the paradoxical element of being simultaneously whole "unto-herself and receptive." The Kore is at once impermeable and permeable, invisible and visible—and through this paradox the Kore possesses a truly creative power.

Persephone is not the only Kore in Greek mythology; by definition Artemis and Athena would also be considered Kores. However, Persephone is unique because she *does* marry Pluto, she joins in union with a masculine principle. She thus possesses a "polymorphic quality," which "comes from her ability to stand in herself."[22] Persephone does not bewail her fate indefinitely, nor does she blindly deny that her world and life have indelibly changed. Instead, Persephone displays what Rossi calls "the most critical feature of the Kore, which is an integrity that transcends phases of life."[23]

Persephone's cyclical descent into and emergence from the underworld also played a critical part of the ceremonies that took place during the Eleusinian Mysteries, which were the greatest and most famous of the secret mystery religions of ancient Greece. Rooted in ceremony, what was taught during the festival of the Eleusinian Mysteries has been debated for centuries, but undoubtedly concerned human and cosmic death and rebirth. While no one knows for certain what occurred during the secret rites, we do know that the mysteries spoke to the *generosity of renewal.* It is commonly believed that during the most sacred rites of initiation, there was a reenactment of the abduction of Persephone and her return to her mother Demeter, which was watched by the initiates, who had fasted in preparation, and then drank a concoction called *kykeon,* which evidence suggests was hallucinogenic. One of the final rituals is thought to have included the use of an ear of grain, "which was understood to be Persephone rising form the underworld."[24] Whether this was meant to be symbolic or indicative of a true transmogrification is unclear, but the ceremony is thought to have comforted and inspired initiates by showing that life not only descends into death, but rises from it as well—thereby revealing the end and beginning of life to be joined in a single locus.

It could certainly be argued that we in the West value technological progress over human life. The history of nuclear weapons and the unfolding present of artificial intelligence makes this perfectly clear. Perhaps we truly are no longer in the domain of the Great Mother—the current state of our environment and climate certainly point to a great divorce from her principles. It seems we may have been led far astray by a destructive masculine energy, disguised as progress, which may not be containable or de-invented. Perhaps the history of Pluto, and its correspondences to our situation today, tell us that we have to discover a new feminine archetype to turn to for renewal; one that is whole unto herself, capable of graciousness, mercy and renewal in the midst of unthinkable conditions.

Just as grain absorbs nutrients out of the dark earth through its fibers, so too does Persephone find nourishment in a realm of deep darkness. She is generative—not as a mother, but as a guide for souls, and in her vegetative expression, as wheat that may nourish human bodies. Persephone's story tells us that we can find life, love, grace and mercy within realms that frighten us, but only when we are willing to make choices that reflect the new circumstances in which we now live. Even in the face of the oceanic reality of death, we can—and must—affirm life. However, Persephone also shows us that this can only occur when we step out of denial into a place of honoring death. To honor death is to honor life. Until we are able to do the former, it is unclear if we will be able to do the latter.

How can we learn to locate the inexhaustible life force within a dawning darkness if we live in a culture that sticks its head in the sand? Like Persephone we mourn the lost world. But like Persephone we too must be willing to transition into a new picture of reality. We must enlarge our psychic and social ecosystems to accommodate the truth, and locate new expressions of mercy and renewal. We must develop maturity and be willing to recognize cycles of life and death that extend far beyond our own mortality. We must learn what it means to traverse an old world and a new world at the same time. We must go down to the very foundations of life and begin to reorder ourselves from there. This is a difficult transition to endure, but the myth of Pluto and Persephone tells us that it is the acceptance of the inalterable which actually leads back to wholeness—perhaps (or almost certainly) a greater wholeness that was previously possessed.

Make no mistake—the stakes are as large as life itself. Our willingness for adaptation and response must be just as large. If we stay asleep to the changes afoot, we will lose whatever potential for life and love that might exist upon awakening. A new path must be forged that incorporates the real terms of our evolving existence, but the integration cannot take place until both dimensions—life and death, the visible and the invisible—are acknowledged and honored. Persephone appears to show us what to do in the face of great overwhelm and fear: become queenlike, develop mercy, constancy and responsibility, move into a rhythm of disciplined dynamism, be ready for sacrifice, look for life and nourishment in places you might not expect to find it. May we follow her example as we move into whatever lies beyond the black hole.

a new view on synchronicity

NUNC FLUENS FACIT TEMPUS,
NUNC STANS FACIT AETERNITATUM

(The Now that passes produces Time,
the Now that remains produces Eternity)

—Boethius

Man's curiosity searches past and future
And clings to that dimension. But to apprehend
The point of intersection of the timeless
With time, is an occupation for the saint

—T.S. Eliot

What is time? On one hand, time is a puzzling enigma that has eluded all of humanity's attempts at total comprehension. On the other hand, time is the most familiar thing in the world. In fact, "time" is the most commonly used noun in the English language. As humans, we share a particular understanding of time when it is used as a factor in how we organize information—as something that passes, as something that is measured or is used to measure. However, there remains no scientific consensus as to what time really *is*. Time, as an entity, appears to be impossible to pin down. The only other mystery that comes close to time in scope and scale is consciousness, with which time is inextricably braided. Despite its perplexity, time is inarguably a substance that presides over all of human life. Yet the essence of this substance retreats from scientific understanding the further it is pursued. The physicist Carlo Rovelli likens the contemplation of time to examining a snowflake that has landed on your hand: "as you study it, it melts between your fingers and vanishes."[1] Although a total scientific understanding of time remains elusive (for reasons we will discuss later in this essay), the realness of our *experience* of time—as something that exists and flows—is a self-evident truth.

The status of the scientific paradigm as the primary method through which reality is deciphered is a relatively recent development in human history. Mythology, in all of its complexity, is the original narrative method humans used to describe their world, and is one of the great foundations of human thought. Although the threads comprising any mythology often intertwine (making myth difficult to codify) the psychological value of a myth may be evaluated in light of the inner experience of meaning that the myth provides. Carl Jung considered myths to be the symbolic results of the mind. As such, myths do not necessarily have to speak to an objective outer reality. Instead, myths can refer to abstract, emotional, conceptual or internal realities. Time, like most universal human experiences, is expressed within mythologies all over the world, and in fact, every culture that has a concept of time also has a mythology of time.

Because the riddle of time has never been fully solved, its study remains open to all investigative perspectives. Reality has often proven itself to be quite different from what we instinctually believe it to be, and this is particularly true for time, because the more we examine time's phenomena, the farther we end up from certainty. The disorienting depth of time's mystery is woven so deeply into our humanness that both rational scientific theory and imaginative conjecture are equally valid methods of pursuing a definition of time.

One of the most cited—and still relevant—musings on time is from the third century CE, and comes from the Christian philosopher Saint Augustine, who asked,

> What then is time?
> If no one asks me, I know what it is.
> If I wish to explain it to him who asks,
> I do not know.

Augustine understood that something about time is linked to human knowing, but also that this link between time and knowing calls our knowing into question. How exactly can we know what we know if one of the vital elements of our reality is a shape-shifting enigma? Time rules over humankind's perception of all phenomena and yet the phenomena of time can never be fully perceived.

The more that the question of time is pursued, the more the structures we depend on to define reality—past and future, subjectivity and objectivity, inside and outside—crumble. But the beauty of engaging with these kinds of mysteries has to do with this very deconstruction, and with the inner experience of one's existence that such an endeavor of thought provides. Time is at once both extremely intimate and undeniably universal, and it is from its universal structures that our individual experiences of time form. In this way, the experience of time's passing is so ubiquitous that it is not only foundational to our concept of reality, but often seems synonymous with life itself.

The extent of time's phenomena is truly inexhaustible, and because of this, the pursuit of time's truth remains an animated playground for some of life's greatest attempts at self-definition. In this essay, we will juxtapose modern Western scientific and philosophical perspectives on time—drawing from arguments made by the physicists Carlo Rovelli, Brian Greene and the neuroscientist Dean Buonomano—with Greek mythological and classical Chinese philosophical and mythological perspectives on time. The conversation between these differing views will then be used to investigate the phenomenon of synchronicity, as defined by the analytical psychologist Carl Jung and the physicist Wolfgang Pauli during their lengthy correspondence. What will hopefully emerge from this endeavor is a new, dynamic portrait of time—and synchronicity's place within it.

There are two major philosophical theories on the nature of time: presentism and eternalism. Presentism asserts that only the present (the "now") is real, while eternalism, in contrast, asserts that the future and the past are as real as the present. Through the lens of presentism, the "past" describes a specific arrangement of the universe that no longer exists, while the "future" describes a configuration of the universe that has "yet-to-be-determined."[2] This is the experience of reality that we as humans know and share with one another. The past is gone and the future has not yet arrived. We live in a perpetual state of "now," and somehow these "nows" accumulate into a past that we can remember but cannot access in any physical way. Similarly, we can imagine the future with excitement or dread, but that excitement or dread exists now. These feelings do not actually belong to the future, they belong to the present because we experience them in the present.

PRESENTISM

The eternalist paradigm, on the other hand, asserts that the future and the past are as real as the present. While presentism would say that the "now" is the only thing that exists, eternalism considers the concept of "now" to be a lot like the concept of "here." Although a subject may be at a specific point in space, there are other real points in space that exist where that subject is not. Similarly, the model of eternalism tells us that, although a subject may exist in/at a specific point in time, there are equally real moments in the future and in the past in which older and younger versions of that subject (as well as other beings) also exist.

ETERNALISM

Physics, the branch of science most often concerned with the nature of time, operates under (and within) an eternalist paradigm. This acceptance of eternalism also usually entails an acceptance of the "block universe" model, discussed below. In order to understand why a branch of science would adopt a model of time that goes completely against our intuition and lived experience of how time works, it helps to understand Albert Einstein's theories of general and special relativity. And in order to understand Einstein's relativity, it is necessary to take a quick look at the ideas of Aristotle and Isaac Newton.

It is often claimed that Aristotle was the first Greek philosopher who ever attempted a succinct definition of time. Aristotle considered the concept of change to be fundamental to the nature of reality. According to the philosopher, the natural world was subject to continuous, unending transformation, and time itself was the measurement of this transformation. In his book *The Order of Time*, the physicist Carlo Rovelli explains that this concept of Aristotle's is valid because time is always, in essence, a question of "when." For instance, if on the night of a new moon I were to ask you, "When will the next full moon occur?" your answer would be "In fifteen days" (14.8 really, but we're rounding up). This answer is a measurement of transformation. What you are really saying when you say "The moon will be full in fifteen days" is that "The moon will be full after the sun has come up and gone down fifteen times."

Aristotle believed that time and change were linked, so if nothing changes, time does not pass. In this way too, time was understood as existing relative to motion (as in the motion inherent in any kind of change or transformation), but time itself was not thought of as a kind of motion. Aristotle considered "motion" to be diversiform. There are many different ways of moving and many different kinds of movements. He viewed time as a stable entity, even though the existence of time was dependent on motion and the changes that occur within the natural world.

The Western philosophical and scientific view of time followed Aristotle for two thousand years, until the seventeenth century when Isaac Newton arrived at the opposite conclusion. Newton asserted that, in addition to the kind of time that Aristotle spoke about, there was also another, "truer" kind of time, which itself was absolute. This truer time progressed on its own in a consistent uniform flow, independent of any external circumstance or perceiver. So, for Newton, there were two kinds of time. There was the kind of time humans can perceive, which was the kind that Aristotle referred to—Newton called this relative or common time. Then there was absolute time, the "true" kind of time, which humans can only deduce through calculations. This kind of time was autonomous and could be applied uniformly to every point in space. This truer time was somehow separate from our everyday reality, and could only be accessed through mathematics. (It's worth noting that both thinkers regarded space in a similar way as they did time. For Aristotle, space was defined by that which makes it up [air, gases, etc.] while for Newton, space existed on its own, even when nothing was there.)

For a few centuries, science favored Newton's hypotheses. Early modern physics was built upon his ideas that time is an entity that runs in an even and unmovable way, and that we can measure our reality based on its imperturbable flow. This model was successful because, on Earth, Newton's laws *do* work, and work well. These are the laws that describe how airplanes fly or how water flows. However, Newton's laws only work in a very particular province of the universe. As soon as you start talking about reality in terms of very large or very small dimensions—the micro or macroscopic levels of the physical universe—these laws no longer apply.

Albert Einstein did most of his thinking in these non-provincial territories. In the early 1900s, Einstein wanted to better understand the relationship between the sun and the Earth. He was curious as to how the two bodies were able to maintain their connection to one another, without any direct physical contact and without relying on something in between them to uphold their attraction. Einstein imagined that a possible explanation might be that the sun and the Earth don't actually "attract" each other, but instead that the mass of the two bodies might continuously affect and alter the space and time that lie between them. If "nothing" exists between the Earth and the sun except space and time, then maybe it was the modification of space and time itself that was responsible for their behavior. Einstein thus proposed that the sun and the Earth might alter and reshape the space and time around and between them, like a bowling ball placed in the middle of a trampoline would stretch the fibers surrounding it.

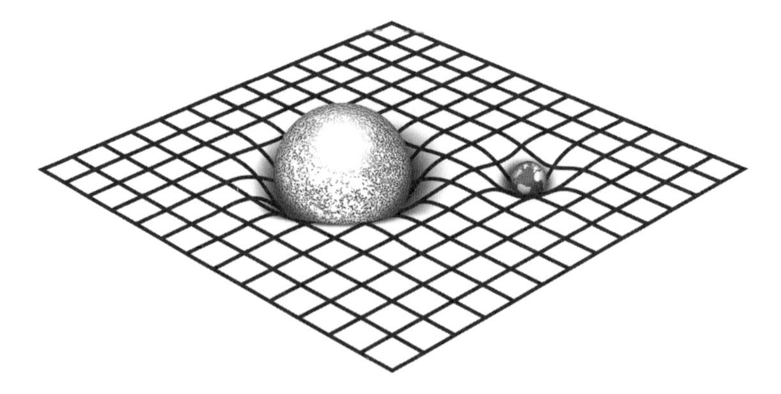

Einstein theorized that this alteration of space and time in turn affects the movement of the sun and the Earth, inducing them to "'fall' toward each other."[3] Then, as the mass of the planet or star sinks into the fabric of spacetime, it slows down the time that surrounds it by stretching out the very structure of time—just like we can imagine would happen with the bowling ball on the trampoline.

This was Einstein's genius. Newton had established a relationship between the force of gravity and mass and distance, but he did not offer an explanation for what gravity really *was*. Einstein offered an answer through his theory of general relativity: gravity is not a force per se, but is instead the very warping of the spacetime fabric itself.

This warping does not just affect the relationship between celestial bodies in space. Rovelli explains that we can see the same phenomenon occurring on Earth, in our relation to one another. Because the Earth has such a large mass, and because its mass slows down the time it sits in, the closer you are to the center of Earth's mass, the slower time will move. Because of this, Rovelli explains, time moves slower at sea level than it does at the top of a mountain, because at sea level, you are closer to the center of Earth's mass. So, if a friend that lives on the top of a mountain meets up with a friend who lives by the ocean, they will each be operating under two unique ideas of what the "correct" time is.

If they each set very sensitive watches at their respective homes and then meet up in the middle, their watches will tell two different times.

If this is the case, is one of their times "truer" than the other? Rovelli explains that no, there is no "truer" value of time. The time that the mountain friend keeps and the time that the sea-level friend keeps only have value "relative to each other."[4] Neither one can be considered more correct. What this means is that our concept of "now" is valid only on a "local" level, and this in turn ruptures our ideas of the "unity" of time. In fact, with sufficiently sensitive instruments, the variance of time that can be seen between the two friends from the mountain and the sea can even be seen between altitudes that lie mere centimeters apart.

The significance of this fact cannot be overstated, because it means that our concept of now is really just a *local phenomenon*. The present is not a constant that extends at full length throughout the universe. Instead, Rovelli likens the present to a bubble that surrounds each of us. The range of this bubble depends on the precision with which time is measured. If we are measuring time by nanoseconds, then "now" can only be determined by feet, but if we are measuring time by milliseconds, the present can be described by thousands of miles.

Rovelli solidifies his argument by explaining that if you took one of the two friends and instantly placed her on a planet four light-years away, there would be "no special moment" on the far-away planet that would correlate with what would be considered "now" for the person remaining on Earth. If the friend on Earth were to look through a telescope at the friend on the other planet, she would see what her friend was doing four years ago, not what she was up to now as understood by the Earth observer.

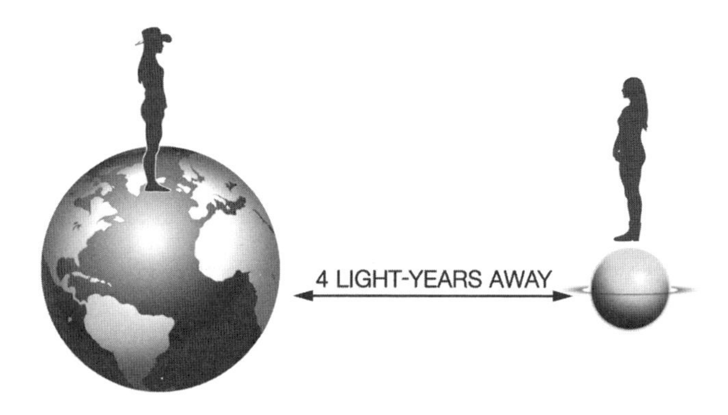

This illustrates that there is in fact no "now" on a universal scale. The notion of the present refers only to things that are near to us. It does not apply to anything beyond a certain local perimeter. This is Einstein's theory of general relativity: Time is slowed down by mass.

Ten years before this discovery, however, Einstein had already figured out that time is also slowed down by speed, passing at different rates according to the speed at which an experiencing subject is moving. This is the theory of special relativity. Einstein essentially concluded that space and time are intimately connected to each other, and that their relationship is such that the "more you have of one, the less you have of the other."[5]

The physicist Brian Greene offers an excellent example of how this can be visualized. Say you're in a car going sixty miles per hour due north. In this situation, all of your energy would be moving in a northward direction:

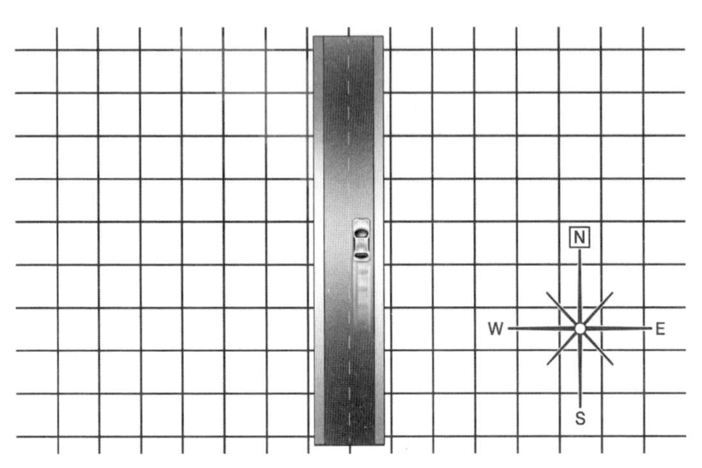

Then let's say you turn onto a road that goes in a northwest direction.

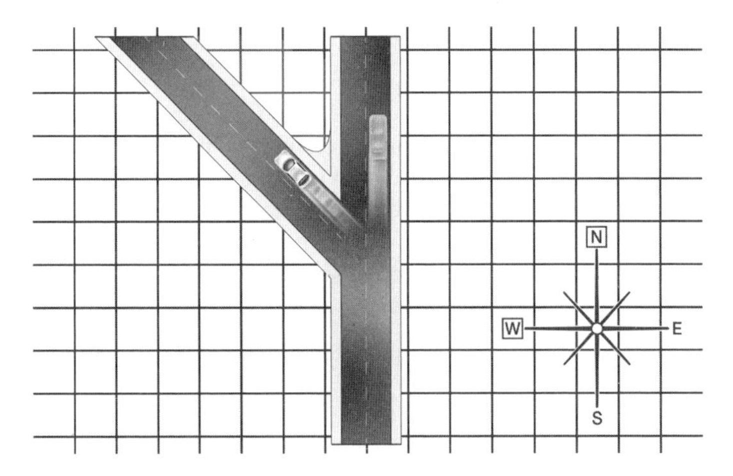

You would still be traveling at sixty miles per hour, only now you would not be making as much headway toward the north as you were when you were on the first road. This is due to the fact that some of your northward motion would now be divided and shared with your westward motion.

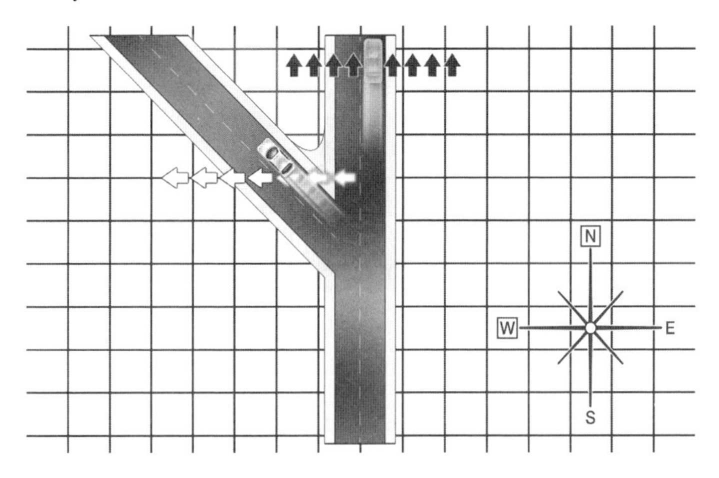

This is because space and time are connected to each other in a similar way that north and west are: If you pull on one, you affect the other.

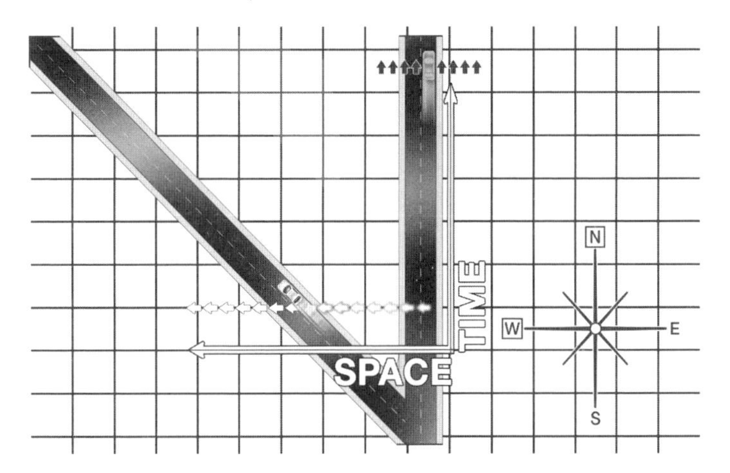

Now what does this have to do with eternalism? Well, because Einstein figured out that motion through space will affect the trajectory of time, this means that they are somehow *profoundly* connected to each other. Working from Einstein's theories, the mathematician Hermann Minkowski (Einstein's former teacher) came to the radical conclusion that space and time were *so* intimately connected that even the concept of space and time being two separate entities was incorrect. Minkowski insisted that even though space and time were relative to one another, only an amalgamation or "marriage" of the two would "preserve an independent reality."[6] This means that because space and time affect each other so much, they have to be bound together in some absolute way. Minkowski built a geometric model out of Einstein's theory that spatialized time into an actual dimension in which the present, past and future are all as real as each other. Through this theory, the universe gets modeled as a four-dimensional "block" with three dimensions for space and one for time. This became known as the block universe.

THE BLOCK UNIVERSE

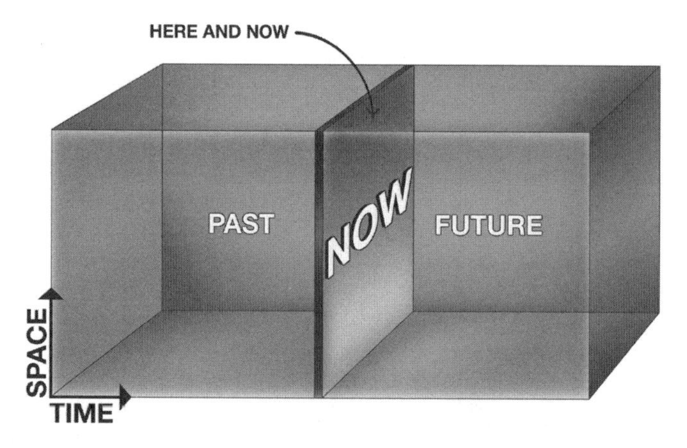

The block universe is inherently eternalist—every event that has ever taken or will ever take place exists indefinitely at a specific point within the block. Here, the notion of time as a fourth dimension is not treated as a mathematical abstraction, but as a physical reality of nature. In this model, the past, present and future are all equal to one another and each is as real as the others. And if every past or future event exists within such a spatialized block, then the issue of "simultaneity," or of differing ideas of what constitutes the present (which is illustrated by the different times displayed on the watches worn by our friends who live on the mountain and by the sea), is resolved. This relativity of simultaneous "nows" becomes solvable in much the same way that two objects in space will appear to be in different spatial relationships depending on where an observer is standing.

Greene has another example that illustrates this well. Let's bring our two friends back into the picture to visualize another way this relationship might be understood. Say we have our same two friends from before—only now they are both back on Earth and one is sitting on a bench while the other is standing 100 feet away from the bench. They are both wearing the same sensitive watches as before.

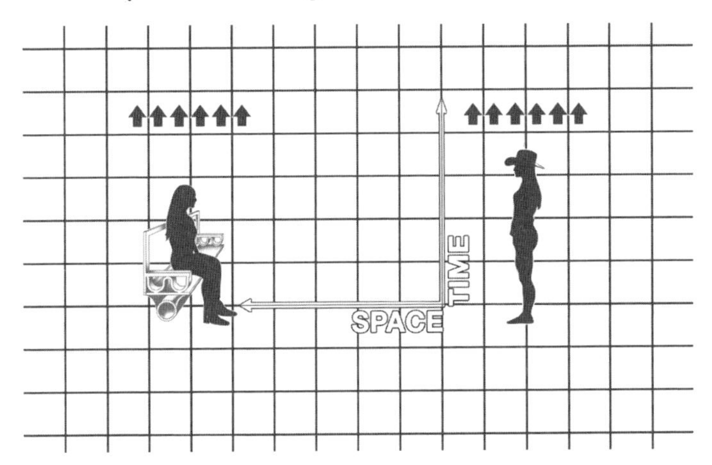

Greene explains that two people holding still in these positions will not be moving through space—but they will be moving through time. Their watches will continue to count off seconds, minutes and hours even if they do not move a muscle. As long as they remain still, Einstein would contend that *all* of their energy is moving through time. However, as soon as the friend who is standing still starts walking toward the bench, the stationary friend on the bench will perceive the walking friend's watch as ticking slower.

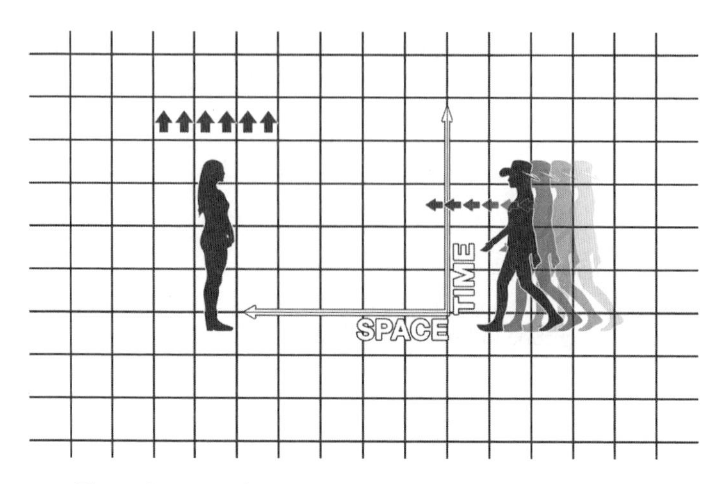

This is because, from the sitting friend's viewpoint, some of the walking friend's energy has been diverted from time into space, in much the same way that the car turning from a northward road to a northwest road has part of its energy diverted from north to west. When the walking friend stops walking, the sitting friend will once again perceive their watches to be ticking at the same rate.

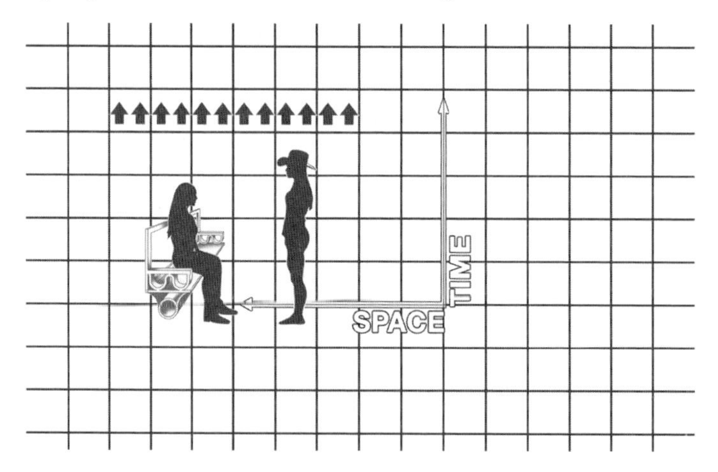

To be clear, the effect being discussed here cannot actually be seen in our everyday life because motion's effect on time is very tiny at the slow speed at which humans on Earth move. However, the principle remains true—space and time are amalgamated in such a way that motion through space affects the passage of time.

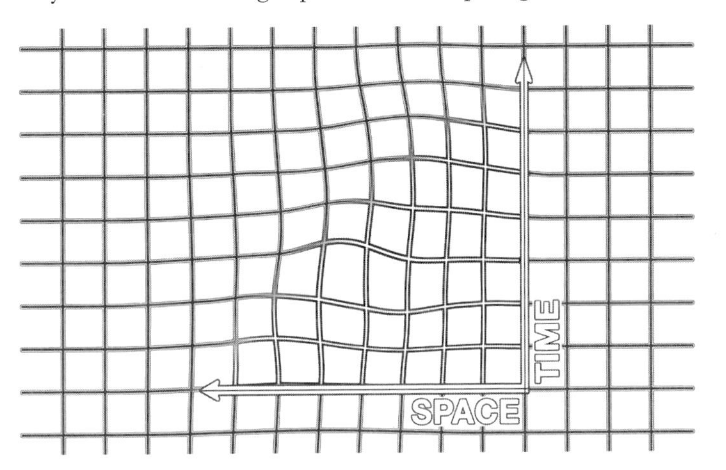

But how might motion's effect on time manifest on a larger scale? To see this, we'll employ one more useful metaphor from Greene. Humans experience time as an uninterrupted flow. For our purposes, for a moment, let's consider that this flow is made up of a succession of moments or "nows." These "nows" line up, one in front of the other, to create a linear series of events. Much like a movie, we experience reality as a continuous stream of impressions, which we could hypothetically break down into a series of still frames or snapshots. Now, if we were to picture that all time throughout the universe operated in this way, we could imagine that the temporal order of the universe was organized as a series of still-frame snapshots, cued up one in front of the other. This would result in a universe that resembles a block, like this:

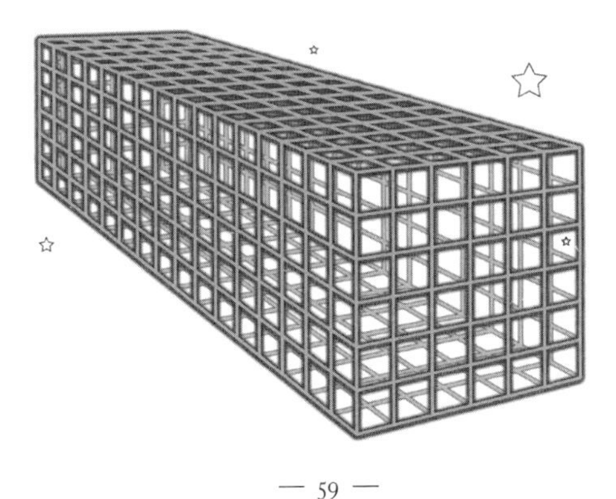

Greene then asks that instead of a block, we might envision this kind of universe as a loaf of bread.

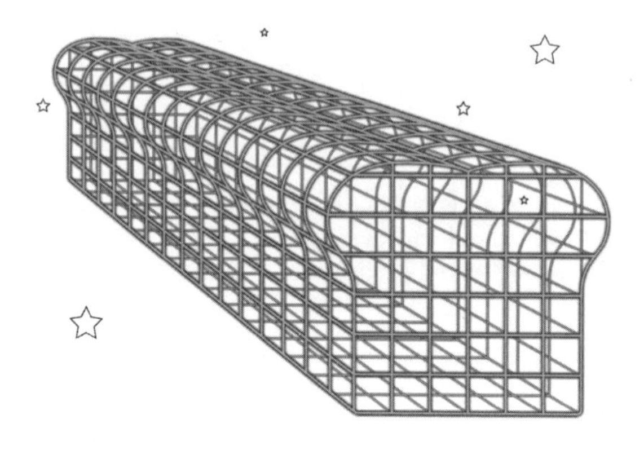

Or, if seen from another angle, like this:

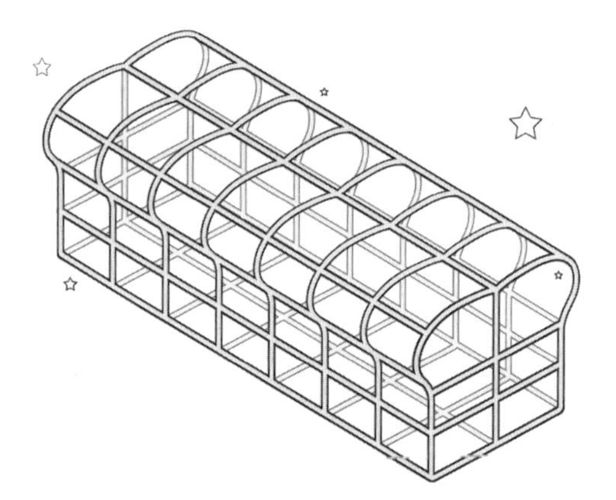

If we then took the idea of time being comprised of snapshot moments and remapped it onto a loaf of bread, we could liken those snapshots to slices of the loaf, "now slices," as Greene calls them.

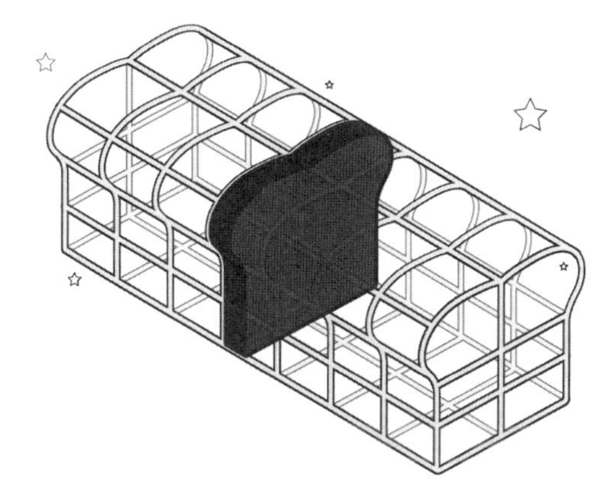

Ignoring for a moment our previous discussion of the non-reality of a universal "now,"[i] we might intuitively conclude that everything happening at the same time throughout the universe (and by everything, I mean everything, even if something is happening in a different region of the universe) shares the same "now slice." We could think of the universe itself as being composed of these slices, lined up from the beginning to the end of time, and anything that occurs at the same moment throughout the universe would share the same "now slice."

However, there is more than one way to cut a loaf of bread into individual slices—

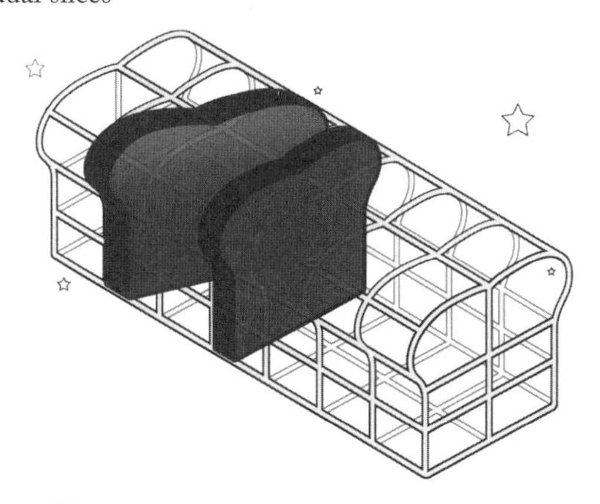

i Einstein developed the theory of special relativity first, so in thinking through these concepts it helps to put aside our understanding of the theory of general relativity for a moment, so we can fully understand where Einstein was coming from when first thinking about these issues.

—and these differently angled slices come into play once you take motion back into account. Because motion through space influences the passage of time, someone who is moving will have a different relationship to time than someone who is standing still. Their "now slice" will be cut at a different angle.

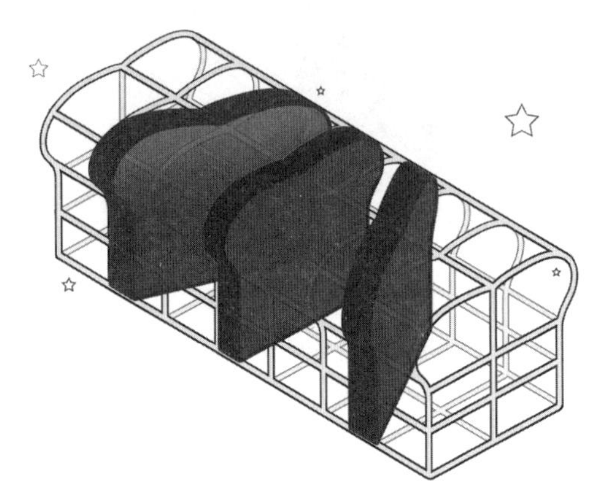

Imagine now that this is the top of the block universe/loaf of bread universe:

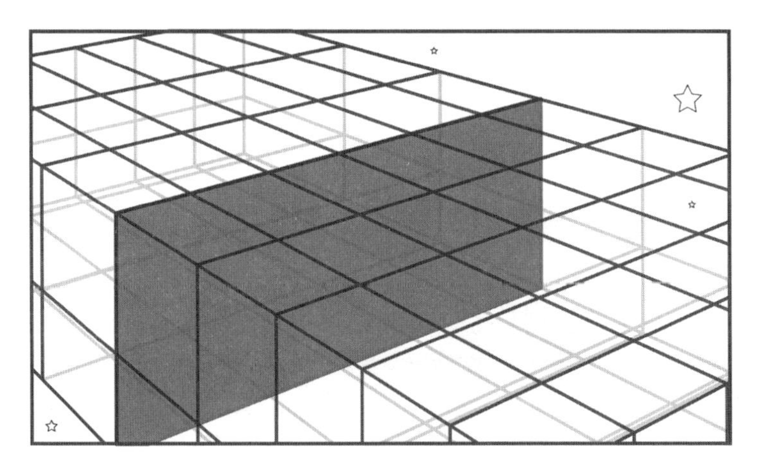

Greene asks us to imagine a person at rest on Earth and an alien sitting in a parked spaceship 10 billion light-years away from Earth. They are both wearing sensitive watches.

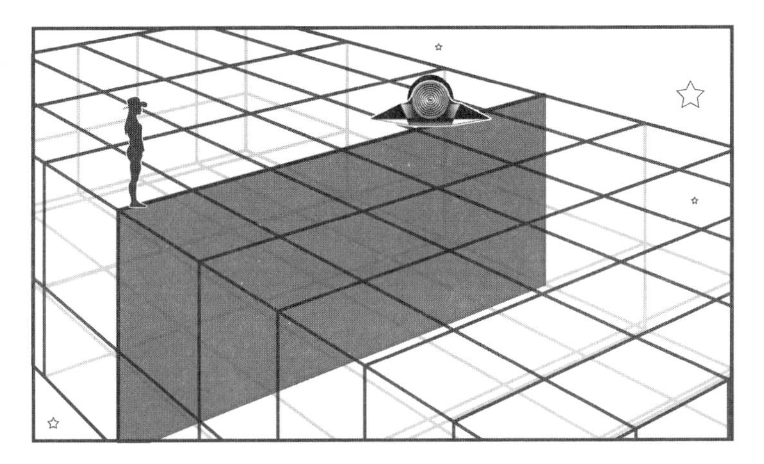

When neither of them are moving, their watches will tick at the same speed. Because their watches are ticking at the same speed, they can be said to share a "now slice." However, as soon as the alien starts to move away from the Earth (even at a very low speed), their watches will no longer agree, and the alien will no longer share a "now slice" with the person at rest on Earth.

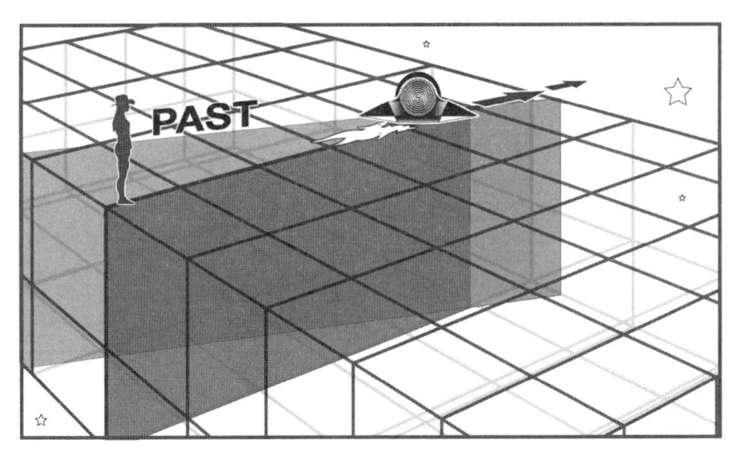

This happens because some of the alien's energy, which was moving entirely in time while they were holding still, has been redirected and is now moving through space (just like the car as it turns from a northward road onto a northwest road). This diversion will cause the alien's "now slice" to be cut into the loaf at a different angle, and this means that instead of continuing to share a "now slice" with the person sitting on Earth, the alien's "now slice" will now be angled toward the past. This is because, if once again we think of the block universe as being comprised of a linear sequence of those film-like snapshots of time, a "now slice" that is cut into the loaf universe at an angle will intersect with events either from the future or the past (relative to the first "now slice").

Now that the alien's "now slice" has a different angle, what the alien considers to be occurring now on Earth will actually be events that happened three hundred years in the past. The alien's new "now slice" will no longer contain our friend sitting on Earth, and will instead include things that happened on Earth three hundred years ago. What is especially remarkable (but makes sense if you think of the absolute marriage of space and time), is that the direction the alien moves in will determine whether or not the angle of the alien's new "now slice" includes events from the past or from the future.

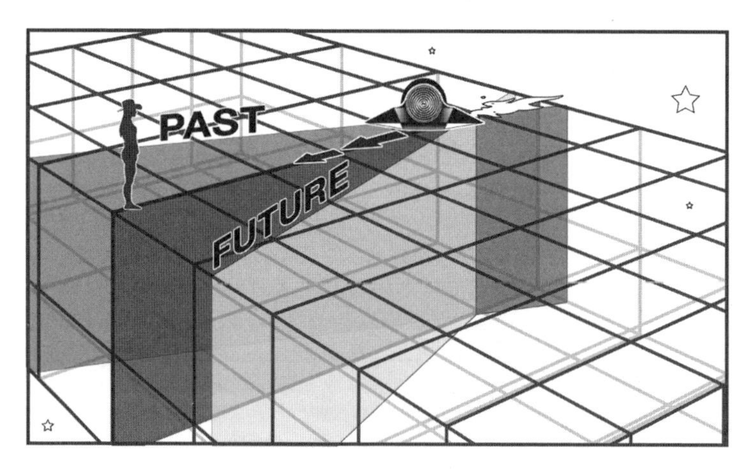

If the alien is moving away from the Earth, as discussed above, then their "now slice" will be angled in such a way as to include the past. But if the alien turns their spaceship around and starts moving toward the Earth, their "now slice" will be angled in such a way as to include events that won't happen on Earth for another three hundred years.

According to the laws of physics described here, the past and future are as real as the present moment. What the moving alien considers to be "now" is not what the person on Earth considers to be "now." If the alien is traveling away from Earth, what they consider to be their "now" will be what the person on Earth considers to be the past, and if the alien is traveling toward Earth, what they consider to be their "now" will be what the person on Earth considers to be the future. If we take a moment to truly digest the way that time is thus affected not only by motion but also by gravity, the sheer awesomeness of the situation becomes clear. Greene sums it up beautifully by explaining that if there is a chance that what you consider to be the past is someone else's present, or what you consider to be the present is someone else's future—and because both experiences of "now" are equally valid (as we know from Rovelli)—then the past, present and future must all be real, they must all exist *right now*.

Einstein's theory of special relativity does imply that all moments in time exist in some permanent location along the temporal dimension of the block universe. And because special relativity has a lot of evidence to support its actuality—for instance the GPS in your car and the clock on your phone would not work if relativistic phenomena were not taken into account within the atomic clocks on the corresponding satellites—there are lots of reasons to accept it as a true, working theory. However, if this is the case, then the block universe (and in turn eternalism) denies one of the most basic features of the human experience: the flow of time. We experience the flow of time as being almost synonymous with consciousness itself. It is so self-evident and fundamental to our being that the suggestion that our universe is actually one big block is not only counterintuitive, but rather astonishing. If all moments in time are real in the same sense that all points/locations in space are real, then the concept of a present that partitions off a past that no longer exists from a future that has not yet occurred doesn't make any sense.

And yet the practical physical applications of this model remain sound, which leaves us with a profound disjuncture between our experience of reality and how reality might actually be organized.

This is where we arrive at the great contradiction: If we live in a block universe, and all moments past, present and future are equally real, then our sense of the flow of time must somehow be a product of our own consciousness. If this is so, then it becomes an issue of untangling a phenomenon that, although originating from the mind, appears to be a feature of the external environment. And when something that is born from the mind appears to be a feature of the external environment, we call this an illusion. Many scientists and philosophers do solve the problem this way, by accepting that our experience of the flow of time is an illusion. If all of time already exists "out there" in the block universe, then time cannot flow, or pass, in the way we think it does. Therefore, a logical conclusion would be to accept our experience of time as a false impression of reality.

When Brian Greene was asked by psychologist Jonathan Schooler about how he reconciles the dynamic quality of the experiential flow of time with a model of the universe where time is merely an illusion, Greene replied, "I see a psychiatrist."[7] He went on to say that consciousness is full of all kinds of illusions, and the flow of time is just one of them.

However, the word "illusion" can and does mean different things to different thinkers in different fields. In his book *Your Brain is a Time Machine*, the neuroscientist Dean Buonomano explains that physicists and neuroscientists often have two very different definitions for an illusion. Buonomano writes that if a physicist were to propose that time's passage is an illusion, they would be proposing that the passage of time exists only in our minds and does not exist as an element of external reality. However, if a neuroscientist were to propose that the subjective experience of time's passing is an illusion, they would be proposing that our sense of the passage of time is a "mental construct," but one that corresponds with or represents (however clumsily) a real phenomenon taking place in the outside world. To the neuroscientist, this notion of a mental construct is correlated with reality: We perceive waves crashing and birds diving into the water because "time is actually flowing."[8] For many physicists, the experience of the passage of time is also a mental construct, but one that does not correspond to anything in the actual physical world.

Clearly these are incompatible visions of nature and reality, but Buonomano explains that neuroscientists like himself do not often have to contend with the debate between eternalism and presentism. For the purposes of their work, "neuroscientists are implicitly presentists." Neuroscientists regard the past, present and future as being constitutionally different because of how the brain functions, which is to say that the brain "makes decisions in the present, based on memories of the past, to enhance [one's] well-being in the future."[9] Buonomano concedes that physicists have good reason to adhere to an eternalist model. On the other hand, however, he explains that neuroscientists ask a whole different set of questions—ones that have to do with the subjective experience of time's passing and the ways in which the brain tells time. From a neuroscientific perspective, the brain's job is to predict the future in order to increase chances of survival, and by performing this very function, the brain operates within a presentist paradigm.

To explain the ways in which time might be considered an illusion of the neuroscientific kind, Buonomano turns to the concept of *qualia*. Qualia are individual instances of subjective experiences, which are adaptive illusions of the brain, but correlate to real physical phenomena that occur in the outside world. Examples of qualia are the pain of a headache that only the subject feels, the blue of the sky as it appears to an individual at midday, or the taste of a particular apple at the moment a single person bites into it. These instances are illusions in the neuroscientific sense because they do not occur in the outside world; they occur within one person's subjective experience. However, they are also adaptive, because each experience of qualia corresponds with some actual external physical phenomena. Certain patterns of waves in the gases that make up the air around us will be experienced as sound or as music, and different chemical combinations of molecules in the air will be experienced as different odors. But there is no inherent quality in any of these phenomena. The radiation of electromagnetic wavelengths correlates with experiences of color, but there is no intrinsic color in any specific electromagnetic radiation. However, even though there is no innate red-ness to the electromagnetic radiation of 690 nm, whenever a person with typical vision encounters that wavelength in the outside world, they will see the color red.

Could our subjective feeling of the passage of time be a similar kind of illusion? And if that is the case, then how did we come to understand what time was in the first place? The prevailing theory in neuroscience is that, in order to comprehend and conceptualize time, the brain probably borrowed and reused the neural circuits that had already evolved to understand and navigate space. Buonomano explains that the brain is very opportunistic and is constantly recycling and co-opting already existing functions and features. Even though space requires three values to pinpoint a specific location, and time only requires one, time is actually much more difficult for the human brain to conceptualize. Mammalian brains have a highly refined "internal map of space" that functions through place cells, neurons that fire when an animal passes through specific points in space. These cells organize themselves into a network that creates a flexible, GPS-like, internal map of the external world. Our senses are involved in this process too—we can deduce how far away something is by hearing the noise it makes and we can use touch to figure out the shape and position of objects. But time is different—animals "cannot physically navigate through time."[10]

Buonomano continues by explaining that although animals track time and engage in prediction, there was probably less evolutionary pressure for animals to conceptualize and represent time than there was for them to understand and map space—meaning it is less likely that understanding the differences between the present, past and future were as critical to survival than understanding the differences between up and down, left and right. Therefore, it is likely that humans developed their concept of time via the brain reusing the neural circuits that were already in place to conceptualize and map space. Furthermore, the way that the brain receives information from the outside world also illustrates a strong correlation between time and space: distance (a spatial dimension) can influence temporal judgement, so the brain uses distances to better make educated estimates about the passage of time.

The depth of the relationship between time and space in the brain can also be seen in language. Almost all cultures around the globe use spatial metaphors when talking about time. We in the West say that a period of time has been long or short, we say "the time is almost here," that someone's birthday is "just around the corner," that a certain meeting has been "pushed up" from Tuesday. However, the linguistic relationship between time and space is not an equal one. We regularly use spatial metaphors to talk about time, but we rarely use temporal metaphors to talk about space. This unequal dependence on spatial metaphors has been suggested as further evidence that human comprehension of time is built on the brain's foundational understanding of space.

If this is true, time as we know it is shaped significantly (and perhaps entirely) by how the brain works. The brain and time are thus inextricably linked. This fact leads Buonomano to ask a strange but important question: If our reality is an eternalist one, with every moment that has ever occurred or will ever occur existing somewhere frozen within the block universe, then our experience of time must be an illusion. But what if the illusion went the other way? Could the structure and operations of the brain have biased our understanding of physical laws instead?

Buonomano explains that, although the brain is a physical organ that must itself obey the laws of physics, our very interpretation of those laws are "filtered by the architecture of the human brain"[11] which performs this interpretation. Because we know that the brain spatializes time, and because the human capacity to ask and answer any question about time is defined by the organ asking the questions, could it be that that scientific models gravitate toward eternalism because it corresponds to the way in which the human brain developed its primary understanding of time? Is it possible that, because the brain conceptualizes time as being space-like, physicists are somehow more comfortable with the model of eternalism (over the model of presentism), even though it goes against our lived experience of how time passes? Considering that science and mathematics is full of models and theories that treat time as a space-like dimension, this certainly seems possible.

Buonomano asks, if the acceptance of the block universe forces us to question whether or not we can trust our own subjective account of something as fundamental and self-evident as the passage of time, is it really so far of a jump to then question whether or not the brain is totally impartial in interpreting the laws of physics themselves? His question seems to lead to more questions rather than answers. Can we ever really know where our minds end and the external world begins?

Will we actually never find the thing we keep looking for, because the thing we keep looking for is the thing that is doing the looking?

Before science became the formal method by which humans describe and analyze the natural world, we deciphered reality through a variety of narrative viewpoints. Mythology, which could be argued to be the original narrative viewpoint, remains the cornerstone for much of human culture and expression. In mythologies from all over the world, time was seen as something mighty and divine. Within these systems, time often possessed one form that was linear and limited. This kind of time was used to describe how things came into existence, behaved while they existed, and ceased to exist. This kind of chronological time became the everyday norm (and the pre-quantum scientific norm). However, this finite form of time was often paired with an infinite counterpart, a timeless time, an immeasurable and inexhaustible form of time to match the finite. As a whole, time was often seen as a dynamic interplay between these two equally important manifestations.

In many ancient cultures there was a direct correlation between the concept of time and that of reality itself,[12] and the image of time as a deity is found all over the world. To the ancient Greeks, time was first mythologized as Chronos, a rather abstract god who was either present at the creation of reality or soon after (Orphic cosmogony tells that Chronos self-generated at the dawn of creation). At some point during antiquity, the primordial god Chronos became identified with the god Kronos, king of the Titans and father of Zeus—although scholars differ greatly on the exact association between the two, and when the amalgamation occurred. Plutarch conflates the two in his essay *On Isis and Osiris*[ii] and Cicero argues in *De Natura Deorum* that the two are synonymous. Of the Titan Kronos, Hesiod writes that he was the "youngest" and "most terrible" child born from the coupling of Earth and Heaven. Chronos/Kronos ate his children, and if we understand Chronos/Kronos to be a symbolic representation of *chrono*logical time, then the eating of his own children is a powerful metaphor for the ways in which

ii In "On Isis and Osiris," Plutarch claims that Cronus was a Greek allegorical name for Chronos.

the linear flow of time takes humans from their birth, through their growth and decay, and finally to their death.

In and after the second century BCE, Chronos/Kronos became linked with Saturn, the Roman god associated with age, restriction and all that time can take away. In his most well-known depictions, Saturn is personified as holding a scythe (a reference to his additional position as a god of agriculture) and an hourglass.

The Greeks also had a second but equally important god of time, Kairos. Whereas Chronos represented quantitative, chronological time, Kairos symbolized a felt, qualitative experience of time. Chronos stood for the seconds that pass, while Kairos stood for moments of opportunity. Kairos was the youngest son of Zeus, and is depicted as a young male god with wings coming out of his back. He is often pictured catching a scale that is hurtling toward him, representing balance and the correct juncture of time.

In his essay "Kairos: The Auspicious Moment," psychoanalyst Harold Kelman describes Kairos as representing a unique meeting point in which a conscious, linear idea of time encounters an unconscious eternal idea of time. The subject to which Kairos appears must be ready to grasp him, because Kairos brings with him opportunities for meaning or wholeness that may not come again. In this way, Kairos must be *experienced*. Kelman writes that these opportune moments cannot be "possessed, saved up and used on the occasion of one's choice."[13] Instead, one must participate totally in the rhythm of life and in the "meaning of existence and of illuminations of wider and wider aspects of a unitive world-view."[14] Kaironic time is in a very real sense meaningful time, which pierces through the fabric of chronological reality, asking its experiencer to answer a divinely determined call.[iii]

At the entrance to the stadium where the very first Olympiad was held in 776 BCE, in the city of Olympia, there were two altars erected. One was dedicated to Hermes, the Greek god of speed and swiftness, and the other to Kairos. Hermes is the Greek god most frequently mentioned throughout the Greek mythological canon, so Kairos's association with him here points to just how significant he was considered to be. Hermes, Eros and Kairos are the only three Greek gods depicted with wings and all three are associated with fate: Eros hits a subject with his arrow, forever changing the course of their life; Hermes is the mercurial agent by which all things change and transform along the lines of destiny; and Kairos must be recognized and seized or one misses the transcendent opportunity he brings with him.

In the Hellenistic period, another time god emerged in which the attributes of Chronos and Kairos were merged—this god was called Aion. Aion represented boundless, infinite time. The concepts of chronological seconds and kaironic moments, as well as all aspects of timelessness, were united under his domain. He encompassed all death, decay, renewal and rebirth, as well as infinity itself. All changes, time-spans, lives, deaths and universes were enclosed within this cosmic symbol. Aion thus represented the utmost principle of dynamic existence.

---

iii Analogous to quantum definitions, Chronos can be seen as representing diachronic time, which is concerned with the way something has developed or evolved through time, and Kairos can be seen as representing synchronic time, which is concerned with something as it exists at one particular moment in time.

Aion is frequently depicted as a man standing in the middle of a circular zodiac with a snake wrapped around him. In many Western mythologies, the snake was often a symbol of time and eternity. This is because of its association with death, its ability to create circular shapes (most obvious when in the form of the ouroboros), thereby representing the cyclical aspect of time, as well as its ability to slough off its own skin, thus renewing its youth. Sometimes Aion was depicted as a serpent hiding its tail under its body, maintaining eternity's secrets. Aion is also represented as a man with a lion's head, who either has a snake wrapped around his body or is holding a snake in each hand (although this image gained more prominence when Aion became associated with the Persian time-god Zurvān).

Aion also originally signified the fluid life-force within all living creatures, which meant that he controlled how long they lived and what happened to them. The Jungian psychoanalyst and scholar Marie-Louis von Franz points out that this association between fluidity and time is also echoed in certain mechanisms used in time measurement, such as water clocks, mercury-based clocks or sand-filled hourglasses.

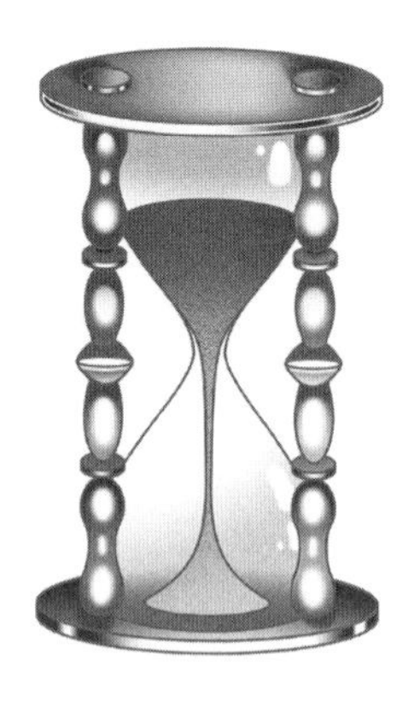

The hourglass, like the one belonging to Chronos/Saturn, is one of the symbols of both finite and infinite time. The sand within the glass runs out, denoting that an interval of time is over. Then, the object is turned upside down and performs the same task, becoming a dynamic representation of a cyclical, potentially unending timekeeping process. As for other timekeeping pieces, von Franz suggests that the archetypal intuition that our linear time is, at its core, cyclical—most probably derived from observing the regular motion of the heavens—could account for the circular shape of clocks themselves.[15]

There was another way in which the idea of cyclical time was incorporated into the Greek mythological pantheon. Among the first godly children that Zeus produced were the Horae, mythological personifications of the seasons, whose name literally translates as "the hours." The Horae were three goddesses who represented the three Greek seasons: Thallo, who stood for blooming, Auxo, who stood for growth, and Carpo, who stood for vegetation. They were the daughters of Zeus and the goddess Themis, who presided over human order, and were attendants of Aphrodite, helping her to maintain the balance and beauty of the natural world. The Horae were often seen dancing in a circle, another reference to the cyclical nature of time's progression.

Within Greek mythology we see multiple interpretations of time as an entity: There is the time that leads us through life to our death, the flow of which no man can stop; there is the time that returns in a cyclical manner, bringing with it at each return an opportunity for new life; and there is an infinite, eternal time that transcends both of these concepts, extending as far into time as it does into timelessness.

⌛

In the mythology of ancient China, time was not personified as such but was seen similarly as a dynamic force that shaped reality. Within classical Chinese philosophy, time (as well as spirit) belonged to the masculine principle of *yang*, while space (and matter within space) belonged to the feminine principle of *yin*. Yang begins all natural processes in the realm of spirit and yin brings those processes to completion in the realm of matter. Yang is creative and yin is receptive. Together, they make up the Tao, which is the guiding law of cosmic order.

Up until the twentieth century, Chinese philosophers believed that reality was not formed by a succession of static states; instead, the world was made up of energy in flux. Accordingly, the flow of time was seen as a product of yang and yin working together in a rhythm. Here, time operates as an activator, carrying what is potential into the realm of what is actual. Classical Chinese philosophy regarded time as a cluster of places and situations, of "coinciding events."[16]

In her book *On Divination and Synchronicity: The Psychology of Meaningful Chance*, von Franz explains that, in a Western view of time, cause always comes before effect—a phenomenon known as causality and represented by the relationship seen here beginning with A and ending with D:

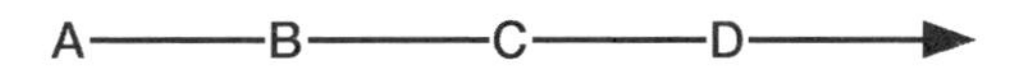

Classical Chinese thought, however, was *a*causal and synchronistic, meaning that events could be linked together through means other than just their physical, causal relation. Instead of "asking" what happened before an event that caused it to happen, von Franz explains that classical Chinese philosophy would instead ask, "What tends to happen together"[17] in time? Time was regarded as a field, at the center of which was the moment in question, surrounded by the events that occurred around it (A, B, C, D, E, F and G).

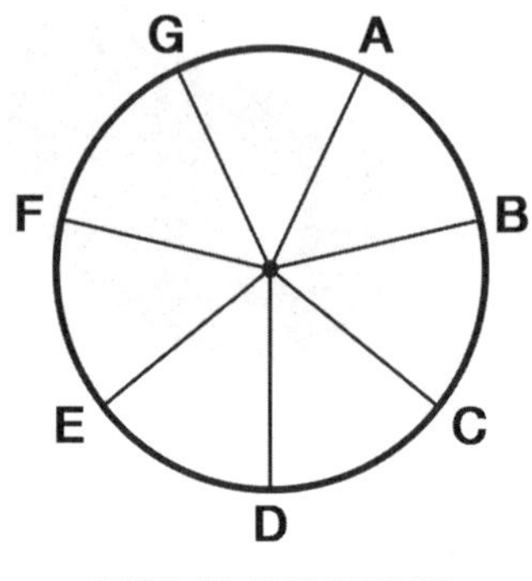

FIELD OF TIME

This way of thinking makes no definite separation between events that take place in the material world and those that take place as felt experiences. The question posed about "what tends to happen together in time" can be satisfactorily answered by both physical and psychological explanations, because the outer and inner dimensions of reality are regarded as equally real and valid. In causal thinking, time exists in the order of "before and after." In this time-as-field school of thought, time is the uniting factor where various inner and outer events collide. The idea of a "moment" then becomes something like a node, the point of convergence through which to observe these two dimensions in their relationship to each other.

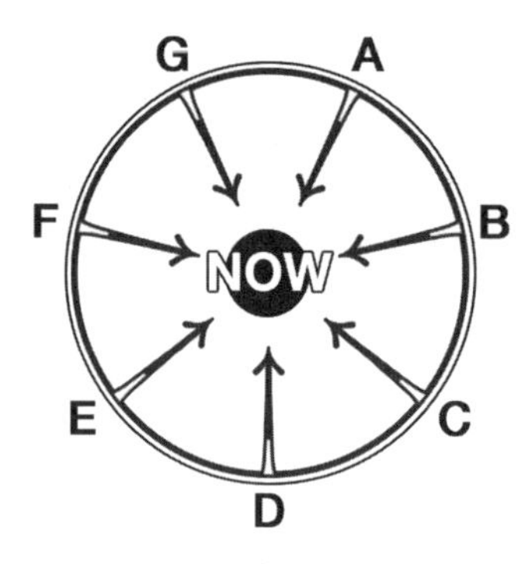

FIELD OF TIME

Within this philosophy are mathematical matrices called number boxes, which act as a kind of algebraic basis for multi-dimensional probability. Classical Chinese cosmology considered the universe to have "a basic numerical rhythm." This "rhythm of all reality"[18] could be expressed in a number pattern. This number pattern mirrored each and every phenomenon that occurred in outer or inner life: All relationships between things and people, and every natural or psychological event, invariably arrived back at the same matrix, which reflected the event itself and the cosmos in which the event took place.

In classical Chinese mythology, there are two major aspects of time, each represented by its own number box. There is cyclical time, represented by the Luoshu, and timeless time, represented by the Hetu.

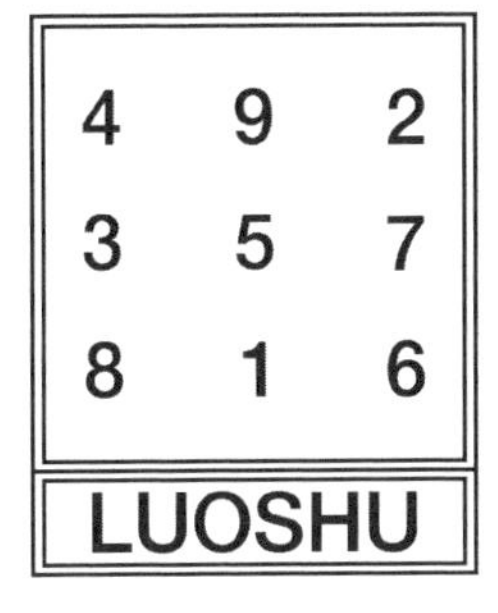

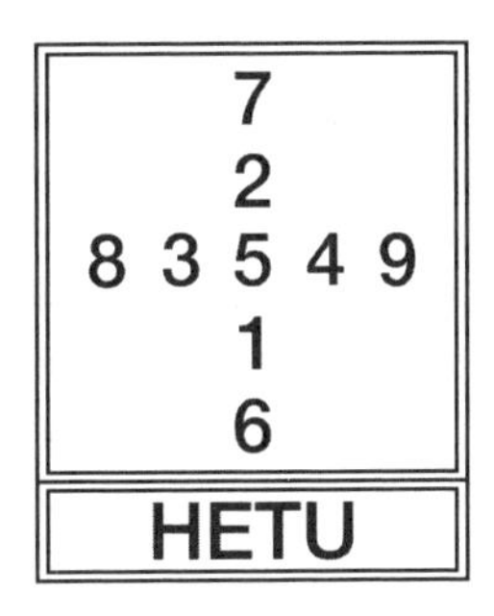

In Greek mythology, ordinary time was thought of as linear. However, in Chinese mythology and classical philosophy, ordinary time was thought to have a cyclical nature, as seen in the recurrent nature of the seasons and the rotating cosmos. Subsequently, human beings live their lives in the realm of cyclical time described by the Luoshu, which could be found at the center of all cultural life. It served as the basis for architectural concepts, the building and tuning of musical instruments and the behaviors of social protocol. It is considered a *magic square* because the sum of all rows, columns and diagonals all add up to fifteen, which was the number of days in each of the twenty-four cycles of the ancient Chinese solar year.

The mythological origins of the Luoshu reach all the way back to the third millennium BCE, when the legendary Sage King Yu, who was known for his ability to control water, staved off a potentially devastating flood. While restraining the waters of the Luo River, a tortoise surfaced from the waves and crawled up onto the bank where Yu was standing. On the tortoise's back were markings representing the numbers of the Luoshu.

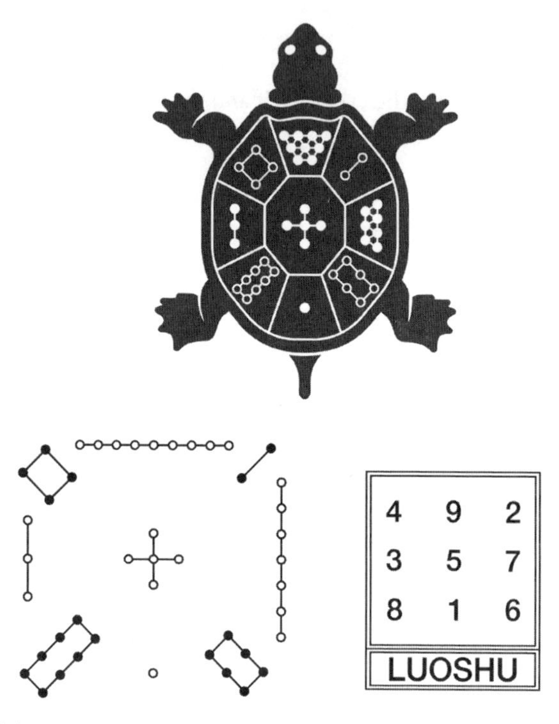

There are nine number cells in the Luoshu square. Odd and even numbers alternate in the perimeter of the box, framing the number five in the middle. The four even numbers occupy the corners and form a cross out of the odd numbers sitting between them. With five in the center position, the sum of any other two numbers connected by the five (for example, opposite corners) adds up to ten. Both nine and five hold special significance within Taoist cosmology. In ancient Chinese numerology, nine represented "completeness, fulfillment, and longevity,"[19] and was used to organize and understand geographic and biological features of reality. The concept of "nineness" was also important to China's understanding of itself in the larger context of the ancient world. The philosopher Zou Yan (ca. 350–270 BCE) wrote that China was one of nine territories in one of nine continents, which were divided by nine oceans, with China inhabiting the middle position. Heaven too was considered to be split into nine regions, with the Divine Rule dwelling in the center region.[20] On the other hand, the number five had a special place in *Wuxing* theory, a metaphysical conceptual scheme that explains a wide range of phenomena, from the movement of the cosmos to the successions of political regimes. Wuxing theory considers Nature to be regulated by five phases, each associated with a fundamental element: wood, fire, earth, metal and water. In these ways and many more, the Luoshu described the dynamic changes reality undergoes in the three-dimensional world.

Interestingly enough, the Luoshu later came to be known as the Magic Square of Saturn in the Western Occult Tradition, and in that school of thought it again represents the basic rhythm of the physical universe.

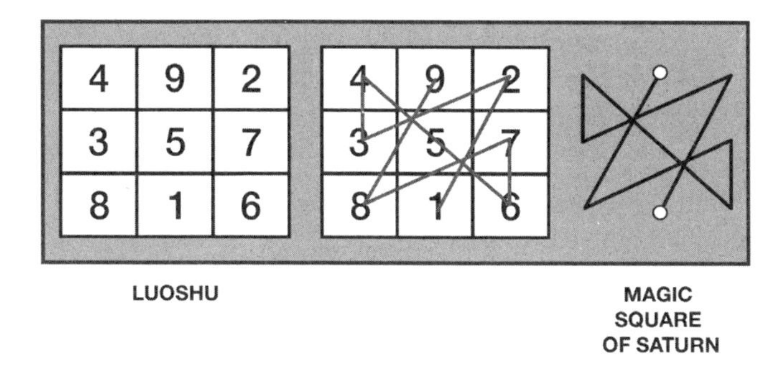

The infinite aspect of time, or timeless time, was represented by the Hetu. Similarly to the Luoshu, the Hetu was also first seen on an animal emerging from the water by a Chinese cultural hero. As the first mythical emperor of China, Fu Xi, was standing on the banks of the Yellow River (of which the Luo River was a tributary), a *longma*—the fabled dragon horse—emerged from the waters with a diagrammatic image of the Hetu on its flank.

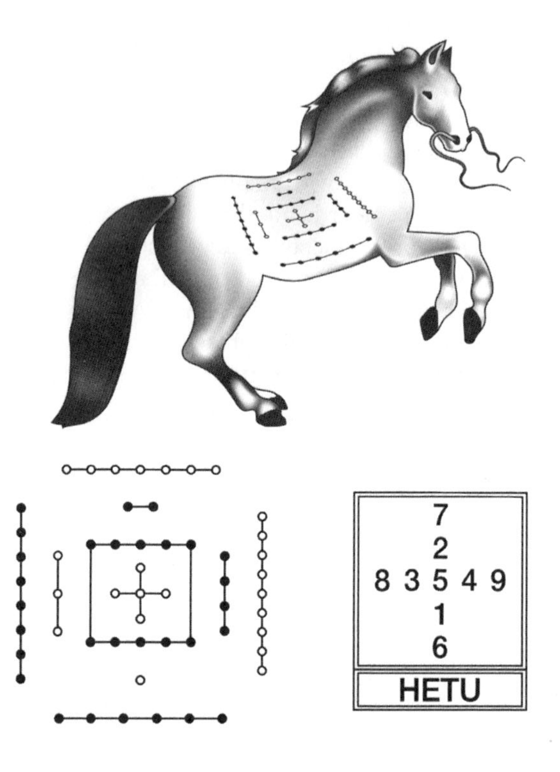

Similar to the Luoshu, the Hetu also has a five (surrounded by a ten, which here is considered to be the higher expression of five) in the middle. But while the Luoshu is comprised of nine cells that form a box, the Hetu makes a cross, and has within it an intrinsic underlying rhythm of continuous movement.

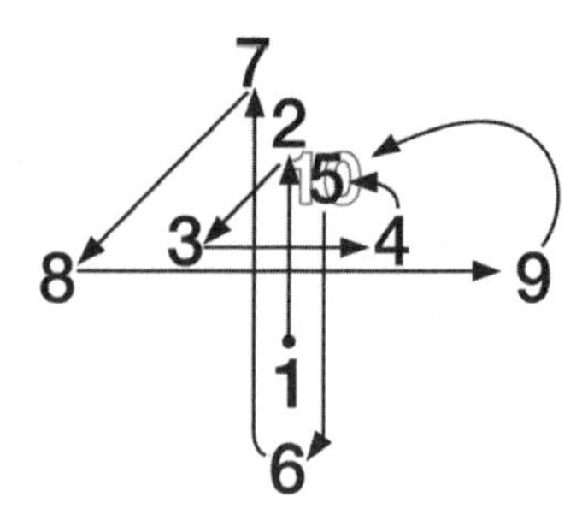

The numbers 1 and 6 sit beneath the 5 in the middle, 2 and 7 sit on top of the 5, three and 8 to the left of the 5, and 4 and 9 to the right of the 5. The rhythm intrinsic within the Hetu follows the numerical order, creating an animated mandala from the movement. Beginning at 1, and moving up to 2, then to the left to 3, then straight across to 4, the movement then curves up and to the left to meet 5 at the center, before beginning again as the movement continues down to 6, up to 7 and so on. When considered three-dimensionally, this pattern, which repeats itself eternally inside the arrangement, radiates out and then contracts back in a systolic and diastolic movement—like a dance.

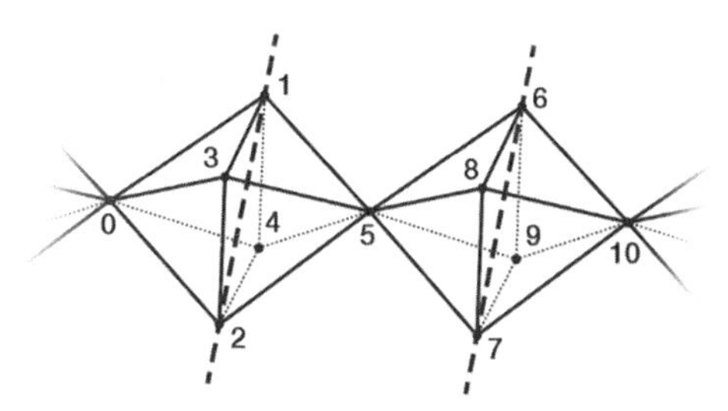

Unlike the Luoshu, the Hetu does not have any linear or cyclical aspects. Instead, its dance-like motion speaks to an ordering of potentiality and power that exists like an undivided timeless continuum beneath our reality. As von Franz describes it, "it forms the static image of a greatly intensified inner dynamism." She also uses the simile of a dragonfly, hovering in one spot midair, supported by the continuous vibrations from its wing movements while still remaining stationary. The Hetu is thus the "primal image of a relatively timeless state of universal orderedness," comprised of "tension and inner vibration" but which, as a whole, is "still and therefore does not enter time or space" without intervention or "interpermeation."[21]

If viewed as a three-dimensional entity, the numerical processions in both the Luoshu and Hetu squares also create a spiraled double helical structure. The movement progression in the Luoshu may be seen as portraying the front view of a double helix while the movement progression in the Hetu may be seen as a top view of a double helix. And as von Franz points out, the archetypal idea of time as a spiral inherently "reconciles the linear and cyclical aspects of time."[22]

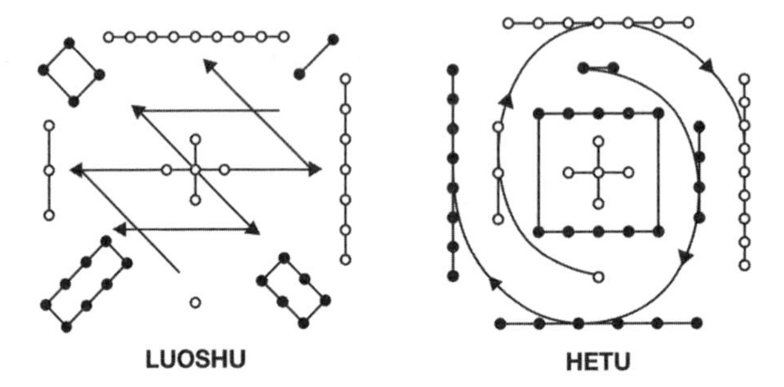

There is yet another structure formed from the movement of the Hetu if viewed from the top. As the movement spirals, it clusters together the odd numbers (which represent yang) through one line of movement, and the even numbers (which represent yin) through the other. As this image swirls together, it creates the yin-yang.

Human beings live their lives according to the cyclical nature of time, as seen in the time-bound realm of the Luoshu. However, according to this system of reality, the cyclical time of the Luoshu is superimposed on the timeless time of the Hetu, which sometimes interferes with it. The importance of chance and acausality in classical Chinese philosophy can be seen in this idea of a timeless aspect of reality permeating a time-bound one. Many of the practices of divination that originated in China, including the use of the *I Ching,* are based upon this concept.[iv]

When we juxtapose the leading scientific and philosophical approaches to the problem of time with the Greek and Chinese mythological representations of time as an entity, a beautiful isomorphism appears. The ideas represented by the Greek god Aion and in the Chinese number box Hetu can be seen as correlating to the model of eternalism. Every moment that has ever existed or will ever exist falls within the domains of all three (eternalism, Aion and Hetu). Perhaps Chronos and the Luoshu can be seen as correlating with presentism. Both mythological representations encompass the passage of time as reality changes from state to state—leading either to a finite end (as in our own death), or to the end of one cycle and the beginning of another (as in the rotating cosmos).

And perhaps Kairos, the god of meaningful moments, can stand for those experiences of significance and awe—the great and pregnant instances of timelessness that we happen upon in our lived experience. These are the experiences of falling in love, of synchronicity, of finding yourself in a numinous state. By having such experiences, we are permitted, however temporarily, to encounter the living connection that exists between all dimensions of time—fleeting moments of eternity, which illuminate a brief understanding of the transcendent within the immanent.

Of these great experiences, perhaps the occurrences of synchronicity tell us most about the nature of time. The word "synchronicity" suggests an episode of meaningful chance. Synchronicities involve circumstances or events that share some kind of common symbolism; that relate to each other through meaning but not necessarily through a material causal connection. For instance, suppose you wake up from a dream in which you have just seen a friend who you haven't seen in waking life for a long time. Later in the

iv The Hetu and Luoshu squares have a direct relationship to the *Bagua* or eight trigrams that form the basis of the *I Ching.* The Bagua are arranged in two different sequences, the "Earlier Heaven" arrangement and the "Later Heaven" arrangement. Hetu corresponds to the "Earlier Heaven" arrangement and Luoshu to the "Later Heaven" arrangement.

day, you're walking down the street when you see a poster in a store window that also reminds you of that same friend, only to turn the corner and run right into them. Or, you're driving on your way to the grocery store and an old song comes on that makes you think of some time you once spent on vacation in Italy. When you get out of your car and go into the market, you hear the same song playing over the store's sound system. When you go to check out up front, the cashier is wearing a T-shirt that says "Italy."

A very famous and often cited example is that of the French poet Émile Deschamps. In 1805, when Deschamps was a young man in Orléans, a fellow he had just met in a restaurant with the last name de Fontgibu bought him some plum pudding. Ten years later, Deschamps went to a restaurant in Paris and tried to order the plum pudding, only to be told the last remaining portion had just been sold to a different customer, who happened to be de Fontgibu. Then, seventeen years after that, Deschamps was having dinner with friends at a different restaurant in Paris and he again ordered the plum pudding. When it arrived, he told his friends all about the curious incident that happened seventeen years prior, and joked that all that was missing from this moment was de Fontgibu. And wouldn't you know, the now-senile de Fontgibu walked into the restaurant, having made a mistake and ending up at the wrong address.

Another famous example comes from the practice of Carl Jung, who coined the term synchronicity. A highly educated and very cerebral woman came to Jung for analysis. She was exceptionally rational, and her excessive rationality was making it very difficult for her to really examine and deal with her emotions in therapy—instead of working through her feelings, she intellectualized them away. Jung was frustrated with her case, and felt that she needed a meaningful experience of irrationality in order to really beget true change. One day, she came to see him after having had a dream the night before in which she was gifted a piece of jewelry with a golden scarab beetle on it. Jung knew that the scarab beetle was a symbolic image of rebirth, and that this dream was indicating that real growth was now a possibility for his patient. Although this understanding had not yet made it into her conscious awareness, it had made it into her subconscious awareness. As she was telling Jung of her dream, he heard something tap at the window. He walked over, opened the window and found that the thing that was tapping was a scarabaeid beetle with a gold-green color. He picked it up, handed it to his patient and remarked, "Here is your scarab." From that moment on the woman's analysis began to progress forward.

Jung used synchronicity "to cover these phenomena, that is, things happening at the same moment as an expression of the same time content."[23] He called synchronicity an "acausal connection principle." As Jung was developing what would become his treatise on the topic, he wanted to create a schema that could express how synchronicities occur within an inter-connected reality. In a letter he wrote in 1930 to the physicist Wolfgang Pauli, with whom he was engaged in a lengthy correspondence, Jung sent a diagram that looked like this:

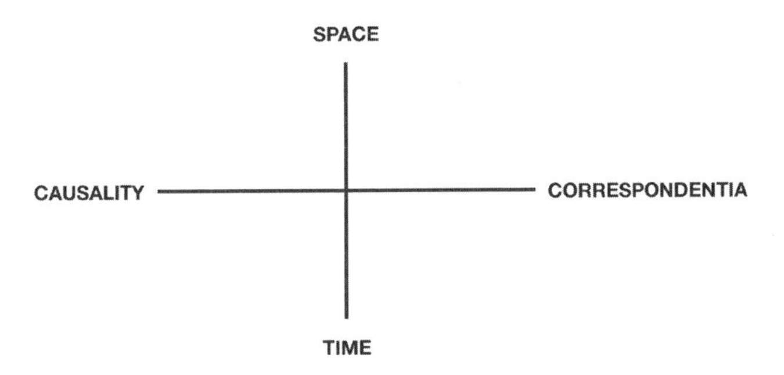

In the schema, space is placed opposite of time, and causality is placed opposite of "correspondentia," a reference to a similar acausal connection principle found in the Hermetic law of correspondence.

In Pauli's response, the physicist pointed out that Einstein's theories of relativity had revealed that space and time could no longer be considered separate entities, let alone opposites. Because, as we discussed earlier, space and time had been merged into a single four-dimensional manifold, Pauli suggested a modification to the diagram, whereby energy (and momentum) sat in opposition to the spacetime continuum, with synchronicity opposing causality. This formulation also reflected Werner Heisenberg's uncertainty principle—that one can never possess complete knowledge of a particle's position in space and its speed/momentum at once. The more information you have about one of the two categories, the less you will have about the other.

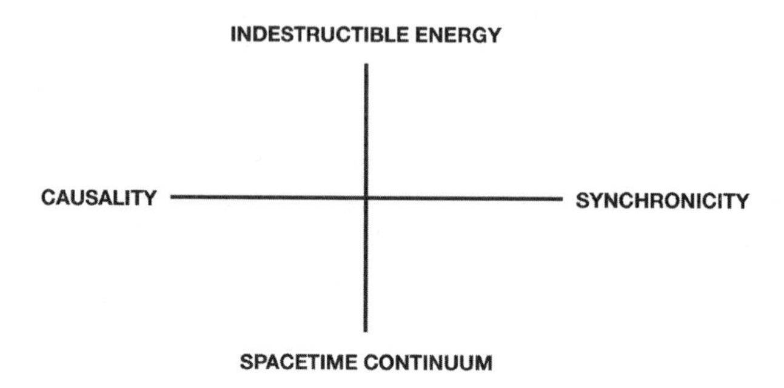

Jung adopted this chart, and since then much of the material about synchronicity written by Jungian scholars has been accompanied by some version of it.

Now if we incorporate the concepts from the Greek and Chinese mythological systems we discussed earlier into this chart, there are apparent connections between the categories in the chart and the phenomena at issue.

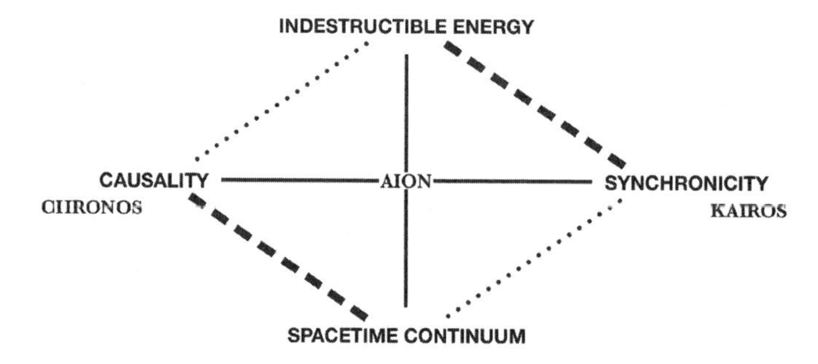

If we first map the Greek mythological perspective of time onto Jung and Pauli's diagram of synchronicity, there seems to be a connection between Chronos and causality, and Kairos and synchronicity. The concepts of succession, limitation and continuation embodied by Chronos line up very well with the idea of causality. In contrast, synchronicity can be seen as those moments of opportunity that Kairos brings with him when he appears, the ones that lift us out of the everyday world of causal relations into a world of deep meaning. Aion, which represents the entirety of dynamic existence, is perhaps best placed at the junction in which these concepts all intersect.

Mapping the Chinese number boxes onto Jung and Pauli's schema reveals a similar correspondence to the diagram's other polarity (indestructible energy ↔ spacetime continuum). The infinite time dimension implied within the concept of "indestructible energy" can correlate to the timeless time of the Hetu, while the Luoshu, the realm that moves forward both linearly and in cycles—like nature and the seasons—can correlate with the realm of our spacetime continuum. In this mash-up, instances of synchronicity can be seen as those moments in which the infinite time of the Hetu penetrates through the Luoshu, interfering with it for brief periods of alignment and meaning.

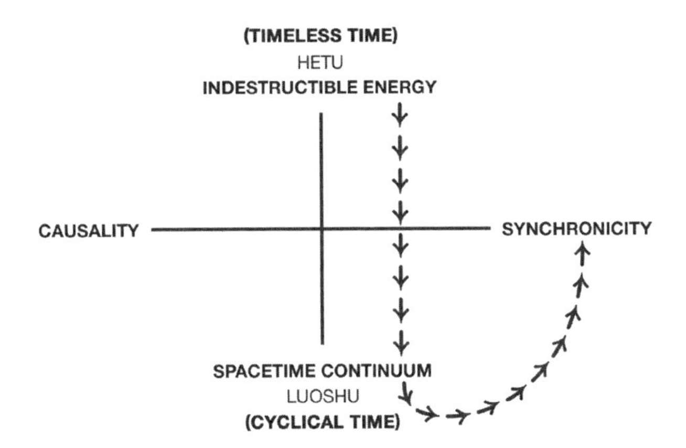

Jung saw synchronicity as an indication of the reciprocal interconnection and interdependence of consciousness and the physical world. As such, experiences of synchronicity can serve as a way of locating ourselves within a much larger and more dynamic frame of time and concept of reality.

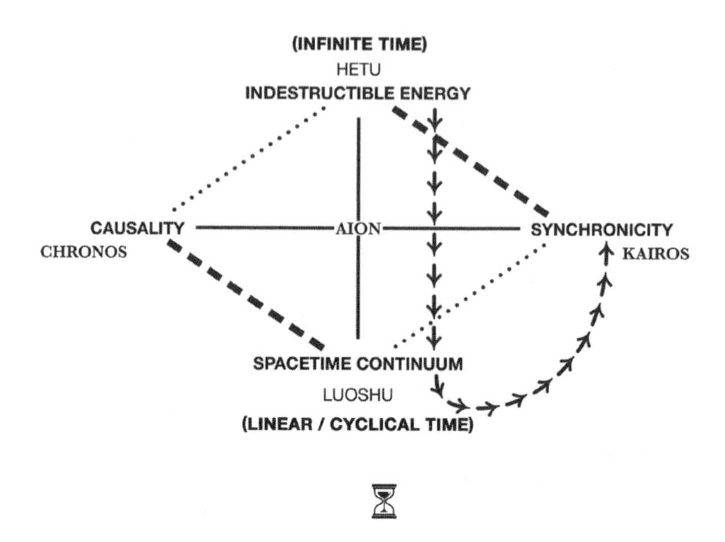

What kind of intellectual care do we give to that which may ultimately be unanswerable? In many real ways, to be human is to *not know*. As the great physicist John Archibald Wheeler put it, "We live on an island surrounded by a sea of ignorance. As our island of knowledge grows, so does the shore of our ignorance."[24] And if our understanding of time is truly based on the brain's recycling of its understanding of space, then how might we ever claim true knowledge of an objective empirical world? Rovelli sums up the problem well when he explains that we are braided together so completely with the reality around us that to imagine a view of "the world" seen from the "outside" makes no sense, because there is no "outside" to the world."[25]

And yet we do catch glimpses of the great mysteries that seem to comprise this "outside" world, and we do have inner experiences, which are truths unto themselves—moments in which time and timelessness seem to hold together within us. These questions remain without answers, but unanswerable questions are still worth pursuing because it is in this Gordian tangle of self and reality that we locate genuine human nature. Even though it is not a picture that can be drawn with definite lines, the inexact image of ourselves we are left with is in fact the most truthful.

Time, for which there is no scientific, philosophical or mythological consensus, does not fit easily into our idea of the world. And yet, without time, there would be no world to conceptualize. We are left with a perpetual uncertainty as to whether or not time's passage is real. But the reality that we live within a mystery, just as it lives within us, remains an eternal truth of the present.

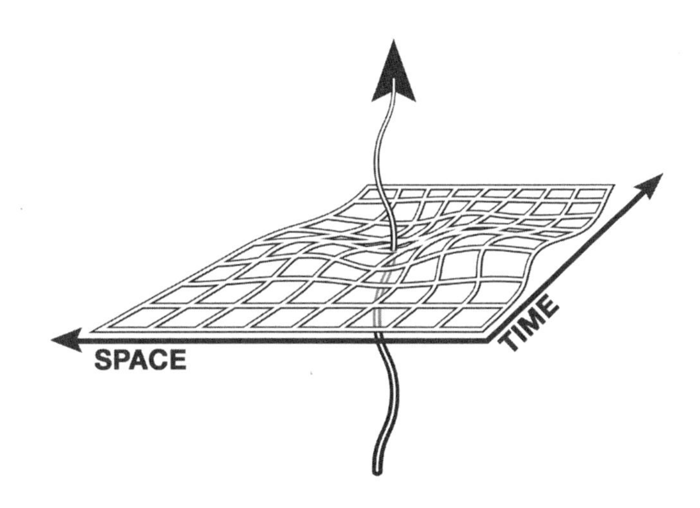

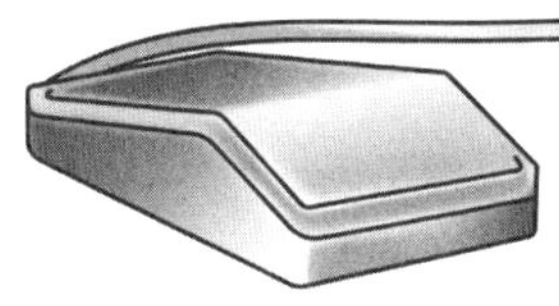

TECHNOMYTHOLOGY

… like a snake changing its skin, the old myth needs to be clothed anew in every renewed age if it is not to lose its therapeutic effect.

—Carl Jung

Are we living in a computer simulation? The question seems to be asked more and more these days. Although the scenario has been well explored in works of fiction, speculation about whether or not we are living in a computer program, run by a more technologically advanced civilization, has crossed over into the realm of serious scholarly study. Notable figures in the fields of science and technology now often weigh in on the subject. Neil deGrasse Tyson has publicly shared his opinion that it's very likely that our world is simulated, and Stephen Hawking thought the possibility was about fifty percent. The American Museum of Natural History has hosted a formal debate on the topic, and Bank of America has gone so far as to issue a report to their clients suggesting that there is a 20 to 50 percent chance that we are living in a simulation. Increasingly, there are scientists from universities all over the world devising experiments to "test" whether or not this prospect is true.

It seems as though references to the "simulation" are riddled throughout popular culture and personal conversations, reinforcing a worldview that many consider to be increasingly plausible. And it might be easy to guess why the idea of a computer-generated reality has taken such a vivid hold of the collective imagination. We live in an increasingly postmodern society where mental, physical and virtual worlds converge and permeate each other. As technology becomes pervasive to the point of ubiquity in everyday life, the psychological and philosophical implications of the human-technology relationship are rapidly evolving to reveal a merging process of an almost unimaginable scope. Advancements in AI have made it so that computer generated images are indistinguishable from images of nature, just as the human experience is synthesizing with digital interfaces. People engage with virtual representations in just about all aspects of life, from banking and medical appointments to social media and online dating. On average, children spend eight or more hours outside of school per day using digital media, and gamers who play massively multiplayer online role-playing games (MMORPGs) average about twenty-one to forty hours a week playing, and that number is growing. As a whole, humanity may already be considered a kind of cyborg, considering our utter dependence on information technology systems: Were they to disappear for any reason, the results would be catastrophic.

At an individual level, the line between user and machine is also rapidly eroding. Cognitively, the brain often does not differentiate between real experiences and virtual ones: The same patterns of neurons fire whether certain representations are real or digital. We export a part of our consciousness into our devices. We house memories in our phone: in pictures, in text conversations, in Google searches. The boundaries between the inner world and the outer world are becoming harder to define as we become more inextricably woven into the fabric of a technological reality.

The technology scholar N. Katherine Hayles coined the term "technogenesis" to describe the relationship that we as humans have with our technologies as we coevolve with them. According to Hayles, these technologies are not external to us. As we invent them and incorporate them into our lives, we fuse with them in increasingly complex ways. We are now beginning to glimpse the personal consequences of such relationships—but what might this perpetual fusing with tech produce on a collective scale? Certainly, just as prolonged interaction with technology alters the personal experience of consciousness, so too must it affect the collective consciousness through the accumulation of such interaction—but what happens at deeper levels of psyche? What activity might this persistent technological integration generate at the stratosphere of the collective unconscious, where archetypes reign and myths are generated? As human consciousness becomes more and more entangled with technology on a collective scale, what might technology's relationship to the symbolic structuring that operates within the collective unconscious become?

We are in the midst of a paradigm shift regarding how we view the world and our consciousness within it, and a shift of this magnitude requires—and determines—a mythological shift as well. If we understand consciousness to be, in part, a story that we tell ourselves, at this moment in time that story must include technology. And if mythology is one way we witness consciousness reveal its agentive nature, which we inherit as a means of understanding ourselves, then a new technomythology must come into being.

Proponents of the simulation hypothesis promote it as a scientific theory, which, if true, would solve age-old scientific and religious mysteries. Some thinkers have accused the hypothesis of being a repackaging of perennial religious structures under a scientific cloak—but the dynamic is a little more complicated than that. It is not as simple as a translation of a certain framework from the language of one discipline to another, although such an analysis is relevant and informative. One way of interpreting the narrative that underpins the simulation hypothesis is to view it as a *creation mythology*, mapped onto the technological possibilities of our time and of the future. It is this tension that makes it unique. It is the same tension that presides over humanity's ever-changing understanding of our current technological reality, as we navigate the quickly closing gap between the "now" and the "what's ahead." As we find ourselves caught somewhere between the natural world and the digital universe, a mythological reading of the simulation hypothesis may reveal psyche's attempt to make meaning of this transition point. If approached as a myth, we may begin to see the simulation hypothesis crystallize as a hybrid phenomenon—not quite science and not quite fiction—which perhaps can tell us as much or more about where we are now, as well as where we are going.

The forms and functions of myth are varied and widespread. The narratives that comprise any mythology may be multiple and intertwining, making myth difficult to codify. One source may tell a myth a certain way, while another tells it a different way—is one version truer than the other? Truth, in terms of mythology, is hard to define in the way we typically understand the concept via our externally oriented intellectual tradition. In terms of the purpose to which it is in service, the trueness of myth may be considered in light of the inner experience of *meaning* that it provides. And the larger truth of mythology may lie in the fact that it remains alive and ever-changing, propagating itself via the needs of both inner and outer realities as the world changes.

The psychologist Carl Jung considered myths to be the symbolic results of the mind. As such, the reality to which a myth refers does not have to be objective. Instead, myths may speak to abstract, emotional, conceptual or internal realities, which provide an instinctive knowledge of life's dimension of timelessness as it is experienced in lived time. In the same way we know that there are more colors in the light spectrum than we are able to see (such as infrared and ultraviolet), and more odors than we are able to smell (such as carbon monoxide), so too are we able to possess a knowledge of other aspects of consciousness that are outside our direct awareness. In this way, myth may be considered vital to the holistic health of the human experience, connecting us with an understanding of psyche's interconnected vastness beyond the scope of our everyday orientation.

There are some mythic structures that are distinct from others. Creation myths are of a unique variety that speak to a specific set of intentions and aims. Unlike hero myths, or local myths of an event or place, creation myths communicate something about the entirety of existence, and the fundamental patterns of which existence is comprised. They contain something more than other myths, something that concerns the mystery of the origin of life, both human and cosmic. Creation myths attempt to formulate an image of where, when and how the primordial beginning occurred. They seek to tell of the transition point where the nothing-that-was-not became the something-that-was. However, because it is impossible to have direct knowledge of the moment of our own creation (being the very thing our existence is built upon), there is a paradox inherent in the endeavor. The exact beginning of human existence is a mystery, and because it is a mystery, the psychologist and author Marie-Louise von Franz writes that the unconscious generates "many models of this event."[1]

These models are organized via the *archetypes*. According to Jung, archetypes are primordial psychic patterns that originate from within the collective unconscious. The collective unconscious is objective and transpersonal, and differs from the personal unconscious in both scope and content. The personal unconscious is contained within one person and is determined by their personal experiences, while the collective unconscious is inherited and shared by all. Jung built his concept of the collective unconscious based on his comparative studies of world religions, in which he saw that corresponding mythic motifs and images were present throughout a multitude of different cultures separated by time and geography. From the presence of these analogous representations, he intimated the existence of certain shared propensities for the structural ordering of images, and came to call these dispositions archetypes. In Greek, the word archetype means "prime imprinter," or "original pattern." Archetypes are both objective universal structures that influence the emergence of specific images into consciousness, as well as dynamic agents that constellate themselves spontaneously according to their own laws. We interact with them as both object and subject. Jung compared the archetype's organizing process in the psychic realm to the role of instinct as a structuring factor of nature in the biological realm.

Archetypes themselves are different from archetypal images. As part of the collective *un*conscious, archetypes are irrepresentable, but the archetypal *image* emerges into consciousness due to "some inner or outer state of need." Von Franz writes that the "disposition" for certain structures are "passed down," and those structures produce "similar images afresh." Regarding any given archetypal structure, "psyche makes use of impressions from the external surroundings for its means of expression."[2] Von Franz also points out that whenever we make contact with the boundaries of the unknown, whenever our knowledge of reality ends, "there we project an archetypal image."[3]

In the process of projection that von Franz is referring to, whenever we encounter the edges of known reality, an archetypal formation occurs to fill in the blanks. Ancient sailing maps used symbols or monsters to represent unknown aspects of certain cartographies. In the middle of the map there would be a rendering of whatever landmass was known, and then toward the edges there would be images of magical beings, like dragons or the ouroboros, illustrating uncharted territories. Interestingly enough, there is a similar phenomenon in science called "the lure of completeness." The mathematician and cosmologist Hermann Bondi pointed out that, historically, whenever we encounter the outermost perimeter of our scientific understanding—of what we can measure—we then conclude that there is nothing else beyond that. We think instead that we have a complete picture of reality within our measured framework. In physics in particular, there was once a tendency to believe that nothing meaningful existed beyond the level of reality that then-current technology could penetrate. Events observed at microphysical levels of reality, in very small dimensions or time frames, which reveal relativistic or quantum phenomena, are still sometimes shrugged off by scientists as something that is strange but does not affect the level of reality in which we live our everyday lives.

However, it is at these levels, where what is known merges with what is unknowable, that archetypes are activated. The scientific tradition freely admits that the truth may change. Theories become obsolete as our tools become more advanced and yield more information. On the contrary, in a mythological system, the older the structure is, the more credence it is often given. In part, mythology deals with a metaphoric reality, one that exists within the mind, while technoscience deals with the outside world, in objective measures and mechanisms. But what happens when these two territories intersect and cross-pollinate? From one perspective, mythology may be considered a language of images, and language itself is a technology. However, we also tend to hold the historical perspective that myth stands as a precursor to science. In the way we once turned to mythology to explain how the world worked on cosmic, cultural, religious, class, and geographical levels, we now turn to science and scientific thinking. In the West, as the Enlightenment flourished, the nineteenth century rational viewpoint was that scientific and technological advancement would progressively eliminate magic and all kinds of non-scientific paradigms. The theory was that people would come to think scientifically, because the aims that magic was employed to accomplish would more surely and efficiently be achieved by science.

But this did not happen. There was never an eradication of non-scientific thinking. In fact, now that we as a species are increasingly thinking and acting by means of technoscientific systems, there is a huge reemergence of the mystical in the very spaces where it was thought to be destroyed. This is seen not only in the way in which technology (especially the Internet) serves as the disseminator of all kinds of knowledge, belief and information—but also the ways in which technology now produces or hosts certain magical and mythological structures.

At the same time, if myths are the symbolic results of the mind, then we must also take into consideration that, as our understanding of our world changes, so too will our *symbols*. The symbol of the king no longer means the same thing it did when it was used in alchemical texts. What was once an image of total and supreme power has been relegated to a performative function in society, and over time the symbol has lost some of its potency. The systems of societal organization also change. The mythologist Joseph Campbell points out that, in part, mythologies have historically functioned as fundamental frames of reference for specific cultures, which reinforce specific sets of values. Up until a certain point in human history, mythologies evolved within boundaried spheres of people, with finite horizons of knowledge and purpose. However, as the world becomes more globalized and connected through its technologies, these boundaries are breaking down. Organized religion, once a territory in which the production of myths flourished, is now collapsing in the West, and globalization is changing religious affiliations and spiritual practices worldwide. The thing that connects humans with one another, and which perhaps provides some understanding of unity and togetherness, is no longer only found within a mosque or temple, but is now also increasingly found through a computer screen. Technology has changed the way we see the world, and the impetus that was once expressed by organized religion to create 'in-groups' and 'out-groups' becomes challenged in a globalized world, one in which a person has access to an enormous amount of information at their fingertips.

Would it not make sense then that we might begin to see mythological structures play out within the technological landscape? If technology has risen up as the entity that provides a sense of unity with the world, then perhaps the mythological function of psyche has now migrated to the realm that presently serves to hold us together. Our desire to make sense of the world we live in does not lessen as time moves forward, but as our reality transforms, the way in which we perform the act of making meaning changes. Von Franz describes the ways that archetypes shift based on the changes in outer conditions, which render certain theories outmoded: as external reality undergoes significant change, then a new or secondary archetypal image is naturally pushed forward into consciousness and constellates another model or idea. This new model is then seen as the "truth," while the older model becomes labeled a projection. So, although archetypal images may change as reality requires, the patterns in which the archetypes arrange themselves remain isomorphic.

Again, if myths are the symbolic results of the mind, then we must also take into consideration that our symbols change as our understanding of our *mind* changes. The feasibility of the simulation hypothesis relies on a very specific 'informatic' view of reality, human consciousness and the mind. Although the concept will be explored in greater depth later in this essay, this informatic view has roots in older materialist ideas of reality, and may be seen as an evolution of earlier established theoretical constructs. What is a significant change—and one that may allow the simulation hypothesis to be considered a viable creation myth—is the unprecedented synthesis of information technology and human consciousness that has been a hallmark of the last fifty years. Von Franz explains that before we can even begin to understand what the content of a creation myth may be telling us, we must also remember that "we cannot speak about any kind of reality except in its form as a content of our consciousness."[4] She goes on to write that one finds "creation myth motifs whenever the unconscious is preparing a basically important progress in consciousness."[5] When consciousness undergoes an enormous change, when older representations no longer hold and the balance between an understanding of the outer world and an inner understanding of reality no longer aligns, new creation myths may emerge from within the collective unconscious as a means of regenerating a cosmic model—not through the process of reconstruction, but through the process of re-creation.

Although there are some lesser-known science fiction narratives that predate it, the most well-known depiction of a computer-simulated reality is undoubtedly *The Matrix*. Released in the spring of 1999, *The Matrix* seemed to further fire the Western technological imagination, which was already burning in anticipation of the coming millennium. The film tells the story of Neo, a hacker who comes to understand that reality, as he knows it, is in fact a computer simulation run by an evil artificial intelligence. Through the help of members of a resistance who know the truth, Neo comes to accept that he is "the One" who can fight back against the AI and bring the Matrix down. In part, *The Matrix* was inspired by the work of science fiction legends William Gibson and Phillip K. Dick, the latter of whom gave a famous speech at a French science fiction convention in 1977 explicitly stating his belief that we are living in a computer simulation. According to Dick, a person may be able to tell that the world is simulated by observing certain alterations in reality by which some variable is changed.

Two years after *The Matrix* was released, the philosopher Nick Bostrom published the first version of his seminal paper, "Are You Living in a Computer Simulation?," the final version of which was published in *The Philosophical Quarterly* in 2003. The paper became very influential and remains the cornerstone of theoretical discourse on the topic of a potential simulated reality. In it, Bostrom lays out what he calls "the simulation argument," which presents the probability of our reality being a simulation based on the truth of one of three propositions. They are:

1) The human species will go extinct before ever developing into a "posthuman" stage in which running detailed ancestor simulations is technologically possible.
2) The human species does arrive at this kind of technological maturity, but for various reasons—perhaps lack of interest or restrictive regulations—does not run historical simulations.
3) We are "almost certainly living in a computer simulation."

Basically, what Bostrom is saying is that either the human race will reach a state of technological advancement where it can run ancestor computer simulations, or it won't. If we do ever reach that state, at any time in the future—no matter how long it takes—then it is very highly likely that we are *already* living inside a simulation. This is because if a civilization ever does reach that point of technological maturity, and if they have an interest in creating historical ancestor simulations, then simulated beings will come to *greatly* outnumber unsimulated beings. After the technology to create high-fidelity simulations (which can host simulated experiences of such detail and magnitude as that of our current reality) is invented, all that would be needed for there to be billions of similar simulations would be more computing power. According to the model Bostrom is thinking within, this kind of steady, exponential growth of computing power is a given, and future civilizations will have access to massive amounts of it. So, if a civilization has reached this stage where it is running ancestor simulations, then it is probably running *billions of them*. That means that the number of simulated humans will vastly outnumber the number of non-simulated humans (or other kinds of beings) that exist outside of the simulation. This being so, it is rational to conclude that we ourselves would be among the simulated beings and not a biological member of the base reality from which the simulations are run.

The idea of becoming "posthuman" is critical to Bostrom's theory. Although thinkers in various fields approach the concept in different ways, for this argument the term "posthuman" refers to a state of development in which the persons or entities within a species have transcended the limits of human mental, physical and biological conditions. The simulated worlds of Bostrom's theory would be posthuman civilizations, although the people living inside them wouldn't necessarily know that. As a philosophical position, posthumanism speaks to the deconstruction of "the human" and treats the notion of "the human" as open and capable of evolving into entirely new forms, mostly through technoscientific methods. Sometimes the term is used interchangeably with "transhumanism," although there are distinct differences between them. Transhumanism also considers the notion of "the human" to be open and evolving, but for transhumanists, the goal is usually human enhancement, not the arrival at a state of existence that is posthuman altogether. Some thinkers have highlighted the difference and describe transhumanism as a *trans*itional state between the human and the posthuman. Transhumanists advocate for the development and implementation of technology aimed at enhancing the human form, the present iteration of which they do not consider final. A posthuman form might be considered whatever a future version might be, without necessarily maintaining a centered focus on "the human" as we know it.

Central to Bostrom's argument is a future moment in time when the technologies needed to create such posthuman realities would exist. Although the term became more widely adopted after the publication of Bostrom's paper—and does not appear within it—the theoretical point of this kind of technological maturity is often called the "singularity." The idea of the singularity can be traced back to the 1950s, when the computer scientist John von Neumann commented that "the ever-accelerating progress of technology" would bring about "some essential singularity in the history of the human race."[6] As it is currently used, the term is often credited to Vernor Vinge, a computer scientist and science fiction writer whose 1993 paper "The Coming Technological Singularity" described a point in time after which technological growth would be uncontrollable and the world as we know it irreversibly changed. The term is now employed as a catchall to describe a stage of technological development where artificial intelligence has reached or surpassed human intelligence, perhaps coagulating into some kind of super-intelligence. It is also used to denote a time when enormous advancements in computing power and virtual reality will make it so that there is no distinction between human and machine, and virtual reality will be indistinguishable from real life.

Although the concept of such a future at one point seemed outlandish, it no longer is so far-fetched. Some futurists defend their belief that we are living in a simulation by referencing the history of video games. *Pong* was invented in 1972 and consisted of two rectangles and a dot, which simulated a two-dimensional game similar to table tennis. Now, video games are three-dimensional, photorealistic representations that have millions of people playing them simultaneously. And video game technology is improving every year. Many believe that this trend of growth will carry on, and soon we will see the development of more sophisticated levels of virtual and augmented reality. If one assumes that the rate of improvement will continue, who knows what technology we will have in fifty years from now, let alone 250.

Much of the basis for this kind of thinking can be traced back to something called Moore's Law. Moore's Law is a prediction made by engineer Gordon Moore in 1965, and revised in 1975 when proven correct, which stated that the number of transistors on a microchip would double approximately every two years. The prediction held true for decades and demonstrated technology's exponential progress in terms of processing power. In 2016, this progress began to slow for the first time since the prediction was made—in large part because there is a fundamental limit to how many transistors can actually fit on a microchip. However, there are parallel trajectories of technological advancement that also describe increasingly rapid growth. Ray Kurzweil, a futurist known for being the unofficial spokesman for the singularity, has charted other comparable graphs of technological advancement. The speed of processors, price of RAM and cost of sequencing DNA all show the same results of exponential growth of power doubling every few years. Kurzweil's 2005 book, *The Singularity is Near*, traced this exponential up-curve through history and concluded that this trend of smaller, faster, smarter machines would eventually result in an artificial general intelligence that surpasses human intelligence. Even if this kind of progress slowed down for whatever reason, he argued that this would not affect whether or not the singularity would occur—it would simply affect *when* it would occur. According to Kurzweil, whenever the singularity is reached (his current prediction is around 2045), this kind of super-intelligence will self-replicate, and then those replications will replicate, resulting in an explosion of artificial intelligence that will render most of our conclusions about the world around us obsolete.

Besides the enormous technological power that would be needed, there is another theoretical presupposition that is essential for Bostrom's simulation, one that he admits would need to prove true in order for his argument to work at all—something called *substrate independence*. Substrate independence is the concept that non-biological/non-carbon-based substrates can host conscious experiences. The theory of substrate independence says that if a system can carry out certain processes based on certain computational structures, then that system would be conscious. It all has to do with how one thinks about consciousness and what one considers to be the mind. If one subscribes to an informatic view of reality, which many transhumanists and posthumanists do, then information is what makes up the universe, and is the fundamental bedrock of reality. Although this is a complex view of the world, its relationship to substrate independence can be illustrated by looking at the transhumanist endeavor of "mind uploading."

There are groups of transhumanists (like the transhumanist religion Terasem) devoted to the future practice of digital immortality through the process of mind uploading. Mind uploading, or uploading consciousness, is the theoretical transmigration of human selfhood onto a computational substrate. The possibility of mind uploading hinges on the informatic view of reality, which considers the mind to be made up of information that is both material and replicable, and that this material may be transferred from one substrate to another, making the mind essentially substrate independent. In this view, the human self (and all that composes it, such as abilities and memories) is essentially constituted by an explicit and unique pattern of atoms in a singular brain, which is the information. Thus, after the human in whom the pattern-of-information-called-the-mind originated has died, the mind may remain in existence as a pattern of information for an indefinite (and potentially unlimited) amount of time on a new substrate (usually a computer or server), barring the destruction of said substrate. Eventually, the mind-as-information could be transferred to a computer programmed to emulate brain processes, or it could be held in reserve until the person could be reconstructed by some kind of super-intelligence (like the intelligence predicted in theories about the singularity).

Bostrom's thinking is adjacent to this kind of informatic view, and has to do with the computational theory of mind, which is a cognitive scientific model that has a symbiotic relationship with computer science. According to this theory, the thing that creates conscious experience is some kind of structural feature of the computation that is being performed by the mind. If consciousness is achieved by some kind of computation, and computers perform computation, then in some significant way brains and computers are alike. For Bostrom, it does not necessarily matter what material underpins the computation, which in accordance with an informatic view would be the processing of a unique pattern of information that makes up a singular mind. Therefore, a computer program that performed the correct kind of computation with the correct kind of information could be considered conscious.

Of course this is only one view of consciousness. Some thinkers, like the physicist Max Tegmark, agree and argue that substrate independence is a natural conclusion, given how other natural processes work. Tegmark says that it is not as though a substrate is totally unnecessary for consciousness, only that the details of what the substrate is made out of doesn't matter much. He turns to sound wave phenomena to illustrate. When someone hears something, it is because they are detecting waves of sound that are produced by molecules ricocheting around in a variety of gases that make up the air. Tegmark points out that there cannot be any sound waves in a gas if there is no gas present, but that it doesn't really matter what kind of gas, or what combination of gases, the sound waves form in.

Others in the field of consciousness science, like the neuroscientist Giulio Tononi, disagree with the notion that the brain and computers are directly comparable. Tononi, who is best known for his work on the Integrated Information Theory of Consciousness (a mathematical model of consciousness which proposes that consciousness is linked to a system's ability to integrate information), essentially asserts that the brain is not a computer and does not function like one. He gives many reasons for this argument, including the facts that the brain has special functions and organizations that do not appear to be consistent with performing computations or following a set of exact instructions; that the brain's billions of connections are not exact and therefore unlike a computer; and that the behavior of neurons in a given brain show incredible microscopic variability even though its overall connective patterns may be describable in more general terms.

For Bostrom, the computation model of mind must be correct in order for substrate independence to be applicable, and for the possibility of a fine grain, high-definition simulation with conscious simulated beings to exist. This, plus the enormous amounts of technological advancement and computing power that technologists and futurologists predict will be available in the future, make the simulation argument somewhat plausible. However, there are certainly many other highly plausible, competing views of consciousness and reality, which if true, would render the simulation theory nothing more than a fable.

☝

Now that we have an overview of what the simulation argument is, and what the necessary future-real conditions are for its existence, our job becomes tracing its mythological reality. There is a distinct difference between the simulation theory and the other theories that have been mentioned in this essay so far (the informatic view of reality and the computational theory of mind). It has a component that the others do not: a narrative. The simulation theory tells a story, and not just any story—an origin story of the entire cosmos. It is, in essence, a cosmology, and every cosmology is a literary enterprise. This is not incompatible with a scientific perspective: Seen from a certain perspective, the big bang is just a story about hot gas.

In some ways, the simulation argument can be understood as the culmination of decades of philosophy, cognitive science and computer science. However, a simultaneous evolution may also be traced from a mythological perspective. As a creation mythology, the simulation hypothesis may be the mythic result produced by the collective unconscious, of consciousness's perpetual fusing with technology. However, in order for the theory to be scientifically plausible, we would have to have a clear understanding of not only what consciousness is, but also how to replicate it. Consciousness is considered the foremost property of embodied living organisms that are embedded in environments, however, there is no authoritative definition for it. Different scientists and researchers define consciousness differently according to varying purposes. We know that consciousness disappears when we fall asleep or go under anesthesia, and then reappears when we wake up or when the drug wears off. What we do not know is how exactly consciousness is related to the material processes of the brain.

There are creation myths that, instead of dealing with the origins of the universe, address the origin of human conscious awareness of reality. In her book *Creation Myths*, von Franz suggests that these particular myths describe the origin of the world as being "completely intertwined and mixed up" with "stories of the preconscious process [of] the origin of human consciousness."[7] These creation myths may indicate the necessary presence of a precognitive understanding that must be in place before we can cross the threshold into consciousness. On this basis, she concludes that these myths present the perspective that before we can understand the origin of the cosmos, we must first begin to know the origin of our knowing. In a creation myth found in the *Brihadaranyaka Upanishad*, from India around 700 or 600 BCE, there is a story about Prajapati, the Lord of Creation, or Brahma the Creator. In it, the beginning is described as absolute nothingness. Then Prajapati or Brahma thinks to himself, "Let me have a Self," and creates the mind before then going on to create the world. In an Australian Aborigine creation story, the beginning is described as a time when everything was still and all the spirits of the Earth were asleep, except the great Father of All Spirits. He awakens the Sun Mother and instructs her to go down to Earth and wake all the sleeping spirits so that she may give them form.

In these myths, the stories of the origin of the cosmos are dependent upon and intertwined with stories of consciousness. There must be some agent who is conscious enough to know something before there can be a wider act of knowing. Although it is certainly formulated differently, the simulation hypothesis also requires an awakening of consciousness. This time, however, the awakening happens when we enter a state of knowledge regarding what consciousness is and how it operates. We must come to understand consciousness as computational in order to live in the reality of the simulation. Whether or not that comes to be is, of course, yet to be seen—but here too there is a function that consciousness must serve. In order for the reality that the simulation hypothesis describes to come to pass, we must also first begin to know the origin of our knowing.

Certainly, the philosophy of mind required for the plausibility of the simulation hypothesis is the product of years of academic discourse on consciousness, cognitive science and computer science. The modern ubiquity of technology also undoubtedly plays a role in the growing acceptance of a computational model as well. All of this can be seen as a linear historical tide that has traveled through time, causing the arrival of this current perspective. However, time itself gets a little murky when it comes to myth, and as a myth, the simulation hypothesis has a very interesting relationship to time.

There is a great difference between how time works in the history and operations of technoscience and how time functions within mythology. In creation myths, time can sometimes be traced back to a different singularity, a vanishing point that separates nothingness from original being. More generally though, time in mythology is a formless, shifting entity. In the Greek mythological canon, for instance, different characters interact with each other in different stories on different timelines. In this way mythology remains evergreen, a mirrored pillar upon which we lean and also search for our reflection. The progression of technoscience, on the other hand, may be viewed as a linear evolution, with each new advancement building on its predecessors. As generations of technology replace each other, there is a straightforward history that can be traced.

However, although there is a causal succession of technological advancement, we are also finding that this advancement is continuously speeding up. We are trying to grapple with the way in which our world is changing, and the speed at which these changes are occurring, a speed that seems to vastly outpace the speed of natural psychological processes. Some of this can be understood if we compare how time works in a digital universe to how time works for us in our world. In our natural world, time is regarded as a continuum, a complement to space. In the digital world, where the same concept of space does not apply, time is measured by *sequence*. The digital universes found in computers are bound by two separate singularities: at the beginning there is T (time)=0 and at the hypothetical end there is T=infinity. Along this spectrum, time acts as a measurable set of individual, sequential steps. According to our perception of time in the natural world, this digital sequence is speeding up as technology advances, making it appear as if time is moving faster. Some argue that there is actually no time at all in a digital universe, because sequence and time are fundamentally different.

As we become more embedded with our technologies, and this disjunction of time becomes more potent, our collective psychological processes may be facing an unprecedented dilemma of understanding. So, a hybrid creation myth appears as a response to restore balance, in the form of the simulation hypothesis—still dependent on future-real scenarios and therefore not yet fully formed, but formed enough to rest our feet on. And it is at this juncture that we actually find ourselves as a species: We are caught between two versions of the world, digital and natural, with two distinct versions of time.

Within the simulation argument itself, time is a protean concept. The hypothesis has to do with the creation of the world as we know it, and in this way, it has to do with the past—but it is a narrative of the past that *relies on the events of the future.* Accordingly, we will only come to know the past by how we come to know the future. Although there are a few variations that are commonly explored, one theory about the civilization that may be running our simulation is that they are our own descendants from thousands of years in the future, observing our history. If this were the case, although we would experience time as linear, time as a larger construct would actually be cyclical. There is a long mythological precedent for the concept of cyclical time, as well as precedents for two kinds of time co-existing at once: a linear variant that sits atop an infinite or cyclical one.[8] In some ways this loop-like formulation recalls the image of the ouroboros that was projected onto uncharted geographies and inscribed on ancient maps to represent the unknown. Perhaps our intuition about the cyclical nature of our reality has been with us for a long, long time. We must go forward to go back, and our existence, in some way, may be both the snake's head and tail at the same time.

Another way the notion of cyclical realities comes into the picture is through the concept of nested simulated universes. Bostrom argues that there is a good chance that our creators were in fact simulated beings themselves (and their creators, and their creators), which would make reality in some ways a recursive spiral. Bostrom points out that it may be possible for simulated civilizations themselves to become posthuman. Although the computers they build in their realities to run their simulations would be "virtual machines," this concept is already supported by practices in computer science—like the programming language JavaScript, which is itself a virtual machine. If we, in this universe, ever get to the point where we are running our own ancestor-simulations, then there will be convincing evidence that the first two propositions of Bostrom's theory are false and that our reality is in fact simulated. Furthermore, Bostrom argues that this may lead us to reasonably suspect that our creators are themselves living within a simulation, and their creators as well. However, if simulated realities create simulations themselves, it would also be logical to imagine that the computing power available to each successive simulation would be less than what was available to its predecessor. This would perhaps result in a majority of simulated realities that do not have the power to create further simulations capable of accommodating conscious beings.

In this descending series of universes, which may result in successively weaker realities, philosopher Eric Steinhart sees a sort of digital, infinite "great chain of being."[9] Philosophically, the concept of a hierarchical chain of all matter and life can be traced back to the theories of Plato and Aristotle. There are also plenty of creation mythologies that show long chains of ancestors and gods who exist and multiply before any creation of our physical reality takes place. These chains of ancestors and gods generate each other, until finally creating the world as we know it. In a Japanese creation mythology from the Shintō tradition, there are innumerable gods or divine spirits called *kami*. The first three kami are born from the primordial formlessness as it begins to separate itself into Sky and Earth. They are genderless and shapeless. Then, as the elements begin separating further, more and more generations of kami are born until two are born that become the creators of the world, Izanagi and Izanami (Izanagi eventually creates the islands of Japan by plunging his staff into the ocean and pulling out island-sized clumps of mud). Izanami and Izanagi in turn create many more kami in the form of elemental and geographical deities. Eventually, they have a daughter who was so beautiful that the gods declared her too precious for this Earth and put her up in the sky to be the sun. Later, she would bear a son who became the emperor of Japan, from whom all subsequent emperors have claimed lineage.

There is a similar chain of genealogy, or birth, of the Greek gods, described by Hesiod in the *Theogony*. The *Theogony* tells of how reality began with the spontaneous generation of four powerful primordial deities, Chaos, Gaia, Tartarus and Eros. From Gaia, the Earth came Uranus, the sky. Then from the coupling of Gaia and Uranus, the Titan Chronos was born. The *Theogony* goes on to describe how Chronos castrated and overthrew Uranus, and then how Chronos's son Zeus overthrew him. From there it details Zeus's offspring, and then their offspring—and how this pantheon of gods established control over the cosmos.

In some Gnostic and Neoplatonic cosmogonies these chains of archetypal entities are replaced by numbers and numerical concepts. First there is the 'one,' and then the creation of reality happens when the 'two' develops out of the 'one,' followed by the 'three' developing out of the 'two,' and the 'four' from the 'three,' and so on. In the creation mythology described by the Gnostic theologian Valentinus, there is a genealogy of personified divine powers, beginning with the Autopater, the self-generated father. Then, from an intrapsychic union of the Autopater with Ennoia, (thought), comes Aletheia (truth) and Anthropos (man). These four recombine in a spiritual quartet that produces another quartet, this time consisting of Anthropos, Ecclesia (the Church), Logos (thought, again), and Zoe (life). Then the Autopater, Anthropos and Ecclesia join in a triad to produce twelve gods, six male and six female, who then go on to generate eight more gods, who go on to generate even more.

Later Gnostic systems influenced by Valentinus conceived of these powers not as personified beings, but as numbers and letters, which generate each other. The idea of numbers and letters being the basis of the universe is not unique to the Christian tradition. Kabbalists believe that the twenty-two letters of the Hebrew alphabet are the body of God and will reveal to their worshipper the way of the Lord. They believe this in part because one can rearrange the twenty-two symbols infinitely, in any combination, to make anything.

When considered in light of the simulation hypothesis, the mythological idea of numbers and letters being the foundation of reality translates almost literally into the concept of computer code. Code is the set of instructions given to a computer about what it should do next. Computer code is made up of a combination of numbers and letters or words, which, when arranged in a certain order, will tell a computer how to behave. Interestingly enough, the historical roots of code itself are tangled with mythological roots.

Binary code, from which all modern computer code evolved, was first described by a mathematician named Gottfried Leibniz in 1689. As he was beginning to develop his system of binary numbers, Leibniz encountered the *I Ching*. The *I Ching*, which translates to

"The Book of Changes," is one of the oldest classic Chinese texts and divinatory systems. It is considered by some to pre-date recorded history, although other traditional Chinese accounts say that it was created around the third millennium BCE and then passed down as an oral tradition. The *I Ching* regards reality as describable through a binary system of yin and yang. Yin, the receptive, which is also referred to as the feminine, is represented by a broken line. Yang, the creative, which is also referred to as the masculine, is represented by an unbroken line. These lines combine in sets of three to make eight trigrams, which are then combined to make sixty-four hexagrams, each describing an archetypal situation of human life.

In the *I Ching*, Leibniz recognized an incredibly sophisticated binary system, which produces a representation of reality. By assigning a zero to the yin line and a one to the yang line, the *I Ching* can be translated into a binary system not unlike the one that Leibniz was developing—some sources even claim that Leibniz read the *I Ching* before beginning to develop his binary system, and was directly inspired by the hexagrams therein. But whether or not he came into contact with the book of changes before he developed his own system, Leibniz was so affected by his encounter with the material that he mentioned the *I Ching* in a 1703 formal paper on binary code.

Through his experience with the *I Ching*, Leibniz came to believe that reality was not to be found in a single source, but rather in the interconnectedness of all things. The creator of the *I Ching* is traditionally considered to be the mythological figure Fu Xi, who in some versions of a Chinese creation myth is credited with creating humanity alongside his sister Nüwa. It is striking to recognize how much influence mythology had on the building blocks of the digital universe. From the developments built on the system Leibniz described, we eventually arrived at a concept of numbers that *do* something, as opposed to numbers that just mean something.

It is also of note that the sixty-four hexagrams in the *I Ching* have often been compared to the sixty four codons of human DNA. If we are living in a simulation, then it seems reasonable to imagine that there would be a similar code animating our world, our selves, and the changes that occur within reality.

☝

The concept that there is a base level of reality, truer than our own, from which our substantive reality emanates is nothing new, and the idea that our reality is in fact some kind of simulation has been a fixture of philosophical and religious thought throughout time. In Western philosophy, Platonic idealism regards individuals as simulacra for pure being, and our world as a reflection of a higher truth. In the simulation theory, there is also a concept that the reality we live in is generated from a truer level of reality, one from which the simulation is run. Although the discourse around different levels of reality has been around for a long time, and has transformed along the lines of human intellectual development, speculation about whether or not our world is a computer-programmed simulation has increased alongside the growing prevalence of artificial intelligence and the advancement of computer technology. The conversation has been particularly present in the Western zeitgeist since the 1990s, when social scientists and natural scientists began using computer simulations and models in their research.

In his book *The Simulation Hypothesis*, the computer scientist and video game designer Rizwan Virk presents a thorough unpacking of various avenues of scholarship that support the belief that our world is a virtual one. He acknowledges that new speculation in physics supports the hypothesis that reality, as we perceive it, emerges from an underlying pattern that we cannot access. Virk builds on Bostrom's theory but develops it in a slightly different direction. Bostrom contends that if our reality is simulated, and therefore the universe we observe is only a small portion of the totality of physical existence, then it is possible that the physics in the universe from which the simulation is run may not be the same as the physics of our world. It is possible that scientists like Copernicus actually discovered the laws of our simulated reality, but these laws may not be the same as those that operate at the more fundamental level of reality. Virk expands this idea, proposing that instead of Bostrom's ancestor simulator, which might be programmed to run without much interference, we may instead be living inside a video game.

In order to support his theory, Virk turns not only to science but also to archetypal patterns in religious mythology. Virk is not the first person to highlight the correlation between the idea of simulated reality and the concept of *māyā* found throughout Dharmic religions, which says that the reality we live within is an illusion. However, Virk's rendition of the simulation hypothesis as a video game adds an interesting dimension to this comparison. Dharmic religions involve concepts of karma and reincarnation. Virk notes that the terminology video games use, such as avatars, multiple lives and quests, intersects quite well with these concepts. In Dharmic religious traditions, humans live their lives out in the illusory world of māyā in order to work through their individual karma. Where a person leaves off in one life, they will pick up in the next. This concept that there is a person, or an entity, who exists outside of the universe in which they participate, but incarnates in it as a character in order to fulfill a set of tasks, fits very well into the model of both reincarnation as seen in Dharmic religions, as well as video games.

Although the formulation of reincarnation differs between Dharmic religions, the mechanism of lessons learned leading to a rebirth is somewhat constant. The concept of karma, Virk points out, can be seen as the information you acquire throughout a lifetime. In a video game, when one lifetime ends, you begin a new one and go back into the reality of the video game again. In a Dharmic religious model, you reenter the reality from which you left, but this time as a different being—although you carry information with you from previous lives. In some Buddhist doctrines, that which reincarnates is not some indestructible soul, as some Hindu traditions state, but is instead some kind of bundle of karma. According to Buddhist philosophy, this bundle is the accumulated cluster of cause and effect that dictates the next rebirth. Virk explains that this would be analogous to a unique set of information belonging to a specific consciousness, like those explored earlier in the informatic view of reality and mind.

Where might this information be tallied and stored? Virk invokes the concept of the Akashic records, which is a spiritual repository of karma that exists outside of the substantive universe. Although it has its roots in Eastern religious thought, the concept Virk refers to is the Western incarnation of the term introduced by the Theosophists. The Sanskrit word at the root of "Akashic" is

Ākāśa, which can be roughly translated into space, atmosphere or sky. The Western equivalent may be close to *aether*, the hypothetical fifth element of the world proposed by Aristotle. In a computer-based metaphor, one can see this concept as some kind of infinite server, or cloud, from which an individual might download the information that would constitute their consciousness.

Virk also points out that the concept of reality being a grand video game aligns well with Western religious traditions. In many interpretations of Abrahamic religions, angels or messengers of God keep track of your behavior and this information determines what happens in the afterlife. Virk refers to the Islamic concept of the "hereafter," which in Arabic is *al-Ākhirah*, which contrasts with the "here and now," or *al-dunyā*. On the Day of Judgment, which is called the day of *Qiyamah*, a person's soul is judged based on their behavior throughout their lifetime. This judgment is based on a "scroll of deeds," which is the record of all things ever done by a specific person. The judgment process involves showing an individual—through some visual mechanism—a "life review" of the effect that their actions had on other people and on the world around them. Virk compares the recording of an individual's behavior in the "scroll of deeds" to a video game character's scorecard, which is stored on a cloud, outside the rendered world of the game. He then compares the process of reviewing one's actions on the Day of Judgment to the replay of recorded gameplay sessions in many MMORPGs, which allows a player to review what went well and what didn't in a given session.

Virk also references Jewish and Christian faiths. He cites Genesis 1:3, which describes the beginning of reality as "God said, 'Let there be light!' and there was light." From there, Genesis says that God created the world in six days and on the seventh day he rested. Virk points out that if our reality was a simulation, that this light could literally be the electromagnetic signals used to "spin-up" a computer program (spinning up is when a disk in a drive speeds up to the proper revolutions per minute for encoding onto or reading from the disk).

Bostrom also admits (as an aside) that there are parallels to be drawn between the simulation argument and various schools of religious thought. Although he considers the simulation argument to be naturalistic, Bostrom concedes that those who inhabit a simulation might consider the posthumans who created the simulation as being godlike. These posthumans would have superior intelligence, the ability to reçord any event that happens and be able to interfere in the workings of our world, which are all attributions one might assign to a deity. However, other simulation thinkers have contested this and maintain that there is no inherent holiness or sanctity in the simulation model. In the version where our descendants are the ones running ancestor simulations, the beings outside who are running the simulation are not gods, but people like us.

Some proponents of the simulation argument insist that the theory not only answers several of the mysteries of religious thought, but also resolves age-old scientific questions, including those about the perplexing nature of reality at the quantum level. Virk highlights the relationship between the way games store and render information and the concept of quantum indeterminacy. Quantum indeterminacy is the idea that a particle can exist in one of multiple states, and that you can't know what state it is in until it is observed; this is according to the most popular interpretation of quantum mechanics—the Copenhagen interpretation—which says the act of observation participates in deciding what state the particle is in.

This process is most often illustrated by the thought experiment of Schrödinger's cat. The experiment is as follows: A cat is sealed in a steel chamber with a bottle of poison and a radioactive substance. If the radioactive substance decays, it will emit a particle that will break the bottle of poison thereby killing the cat. Because the radioactive substance has a fifty-fifty chance of decaying and breaking the bottle, you would not know if the cat was dead or alive until you opened the box and looked at it. So, the cat in the box has a fifty percent chance of being alive and a fifty percent chance of being dead. Of course, there is the argument that the cat is either alive or dead to begin with, and we simply do not know because we have not opened the box yet. However, the Copenhagen interpretation of quantum mechanics says that this is not so. Instead, a quantum system (like Schrödinger's cat) may exist in a *superposition state*, which is when a system has two (or more) different states that could possibly define it, with the potential that it may exist in one of both (or any of those) states. And quantum indeterminacy—according to the Copenhagen interpretation—tells us that it is actually the act of observation that decides what state the system will be in. A superposition is described by a mathematical wave function, and the act of observation somehow causes the wave to collapse into one of however many possible states. So, the cat in the box is both alive and dead until someone observes it, and then the probability inherent in the wave breaks down into a state in which the cat is either dead or alive.

If our world is a simulation, then this collapse does not really happen. Instead, the video game might only render the reality that was being observed by a given player. Virk writes that an important principle of video game development is the optimization of limited resources. A three-dimensional virtual reality game would require enormous computing power if it was run at full scale all the time, so one optimization technique would be the rendering of *only that which is being actively observed*. Similarly, even though MMORPGs have their reality stored in some kind of "master" state on a server, the rendering of what is seen by people playing the game is actually done on each individual computer. This aligns with the Copenhagen interpretation of the principle of quantum indeterminacy, which says that the act of observation in part determines reality. Wherever we look, we find reality, but who is to say reality is rendered where no one is looking at it? Furthermore, in this analogy of how video games store data, we see another example of how information about a world may be stored in a system that exists outside of its substantive reality.

In considering how information may be stored and organized in our universe, there is a line of thought that circles back to the issue of time. The astronomer, computer scientist and technologist Jacques Vallée has done extensive work into what he calls the "physics of information." Vallée approaches the concept of information from the perspective of both a physicist and a computer scientist. He explains that the way we think the universe is organized, in terms of the theory of spacetime, is in fact a "cultural artifact made possible by the invention of graph paper."[10] Spacetime is a mathematical continuum model in which there are three dimensions for space, and one for time, which together form a single four-dimensional manifold. Vallée admits that he has always been uncomfortable with the theory because, although you can move in multiple directions within any of the three dimensions that make up the space portion of the theory, we of course cannot move backwards and forwards in time. In fact, we have no real idea why time passes at all. If dimension X is considered to be up and down, and dimension Y is considered to be forwards and backwards, and dimension Z is considered to be side to side, (all of which you can move in multiple directions within), one can easily visualize why it is so strange that time (T), a dimension in which we cannot move in multiple directions, should be treated the same way as the other dimensions.

Vallée regards the theory of spacetime as a cultural artifact that served the purpose of organizing the amount of information we had access to around the time of its invention. If you have a library with 10,000 volumes, it is very easy to organize it based on a series of coordinates, vertically, horizontally, etc.—you could map it out on graph paper. However, modern libraries are now housed on servers where these kinds of dimensions as we know them do not apply. There is no dimension of time in these libraries. Instead, the information is distributed statistically throughout virtual memory, and then that information is retrieved associatively—as in when you search for a certain word in Google. Vallée explains that computer scientists know that organizing information by space and time is an incredibly inefficient way to store data. He concludes that if we had perhaps "invented the digital computer before graph paper, we might have a very different theory of information today."[11]

If our universe was organized associatively, like a modern computer, and not sequentially, like in the spacetime theory, then events like synchronicities—episodes of meaningful chance where contents of consciousness line up with events in the outer world through signification but not through causal relation—would not be supernatural: They would be a byproduct of how reality is organized. If this were the case, and there is no time dimension, consciousness could be seen as the process "by which informational associations are retrieved and traversed."[12] Consciousness would then be that which generates the illusion of spacetime during the act of traversing associations. If this were the case, then spacetime would be a *pan-mathematical mythology*.

This kind of thinking implies that the universe is organized in a much more complex way than any simulation or video game. It would have to be something much bigger, more advanced and sophisticated than anything our current (or hypothesized future-real) technologies could analogize. However, the conclusions presented here still use the computer and the Internet—and how they are organized—as analogies for the universe. These conclusions are still dependent on an evolving relationship between human consciousness and information technology. And they still rely on a model that treats our perception of reality as emerging from an informational matrix.

As the writer and scholar Erik Davis points out in his book *TechGnosis*, previous worldviews saw the universe in more interconnected ways. These worldviews were animistic and "ecological,"[13] and mythic structures arose from mankind being immersed in the natural world. Animism is a lens of viewing the world as interdependent and alive. The word animism comes from the Latin *anima*, which means 'breath,' 'spirit,' or 'life.' In animistic philosophy, all objects, places and creatures have an element of spirit or consciousness to them. Animistic thought reads a certain kind of enchantment into the world, one by which the spirit that animates human consciousness also animates the entire world. With the simulation hypothesis, we once again arrive at some kind of enchantment of the world, only this time it is a digital enchantment. This time the fire that burns and animates the world is computer code, and this code has something to do with human consciousness.

The historian Frances Yates wrote,

> The basic difference between the attitude of the magician to the world and the attitude of the scientist towards the world is that the former wants to draw the world into himself, whilst the scientist does just the opposite, he externalizes and impersonalizes the world by a movement of will in an entirely opposite direction.[14]

The simulation theory seems to be paradoxically adopting both of these attitudes: We have externalized reality through a materialist, technoscientific worldview that is both computational and informational. At the same time, we have drawn that worldview into ourselves with the belief that we are the product of this technoscientific externalization. We have objectified consciousness, and then concluded that the informational materiality of this objective reality makes up our subjective experience.

As we move into a progressively hybrid existence, one that finds us increasingly separated from the natural world, the mythic structures that used to arise from our relationship with nature must now find a way to navigate the technological transformation of our time. We are not entirely divorced from the environment, and yet we are actively losing our connection with the Earth. Global warming and the philosophical relocation of spirit from nature to machine, amongst other great shifts, illustrate the magnitude of the transition in which humanity finds itself. In many ways, our technologies have uprooted traditional spiritualities in the West, and the structures that used to provide spiritual orientation are breaking down. Organized religion is changing in a rapidly secularizing and globalized world, but people still yearn for deeper meaning. And so there may be new mythologies, ones that are not based on cultural factions or religions and instead reflect back to us our global reality. The simulation hypothesis does this, and even though it is far from an ideal form, it is a start. A fledgling myth.

The concept of the future-real—that which must come to pass in order for the reality that the simulation model proposes to exist—places our current society in a very unique position in time. Myths are often considered to have taken place in a time before time, a fabled kind of time that may be considered the *beginning*. If we place ourselves in the worldview described by the simulation argument, then we are *now* in this fabled time of primordial beginnings. In a way, anything up until the simulation is created could be argued to be pre-history. It is our technological reality now, and how it advances, which will determine how we understand our past. Where we are now—with changing models of time and information, with open concepts of "the human" and "the machine," with cosmologies that possess many mythic structural elements as well as scientific plausibility—would all be part of a swirling, pre-understanding of our reality. This understanding would only crystalize after the technological singularity has been reached. In a way, as seen through the lens of the simulation hypothesis, the events taking place in our current time are the very ones that will establish the world and make it real.

The religious studies professor Jeffrey Kripal writes about "the future of the past," which he calls a "progressive spiraling model of historical thinking and interpretation." The future of the past involves reading events of the past through the lens of the present (or future) or vice versa. This practice can be seen, for example, when conservative Christians read current events like natural disasters as the fulfillment of prophecy, or when a round disc in the sky of a Renaissance religious painting is read as a UFO. Kripal writes that "there is no straight arrow" in this practice, but instead a looping "constant return to the past in order to reassess and recalibrate the present toward a different kind of future."[15] Certainly Rizwan Virk could be charged with performing a future of the past reading in his book *The Simulation Hypothesis*, as compelling as his arguments are. But where Virk ultimately misses the mark is that his arguments attempt to resolve past religious structures by viewing them through the lens of the present or near future, thereby participating in a sort of algebraic rewriting of history, one in which the past (a religious structure, aka variable A) is solved for the present or future answer (the simulation hypothesis, aka variable B). As such, Virk does not acknowledge the kind of looping hermeneutics he participates in.

Instead of being the long-awaited answer to unsolved religious (and scientific) mysteries, what the prevalence of the simulation hypothesis actually does is remind us that a creation myth is necessary for the health of the human species. Mapped onto the possibilities of the present and the near-future, the archetypal structure of a creation myth is born again—a fruitful renewal so that we, as a species, may once again locate ourselves as belonging to the natural universe, even if that natural universe was created by a computer. Jung wrote of how "knowledge of the universal origins builds the bridge between the lost and abandoned world of the past and the still largely inconceivable world of the future." He stressed the importance of this understanding, and the assimilation of this understanding in navigating the human experience that we inherit. Without some knowledge of our universal origins, our rootlessness will only grow and our sense of meaning diminish. A purely technological outlook cannot solve the problems that arise from this unknowing. Jung described a technological or entirely rational viewpoint of history as one-sided, and remarked that "one-sidedness never doubts itself." However, he emphasized that "neither technology nor the dominance of reason can stem the tide of the unconscious, irrational counterforces, which reduce the certitudes of the rational mind to absurdity." He concluded that as rationality lays an increasingly absolute claim to the direction of life, "the more intensely the irrational, or the longing for the irrational, makes itself felt as an unconscious compensation."[16]

Jung also wrote that throughout time "man has always lived with a myth," and that the modern view that man can live with no myth of history is "a disease." He continued, writing that because a "man is not born every day," but instead is "born once in a historical setting, with specific historical qualities," he is "only complete when he has a relation to these things." Jung likened "growing up with no connection to the past" to being born "without eyes and ears." He called this a "mutilation of the human being."[17]

In consideration of Jung's conclusions, perhaps there is some deep insight to be gleaned from the current popularity of the simulation hypothesis. As we continue to feel less and less connected to the natural world we were born into, the simulation hypothesis may serve as an attempt to feel more secured to the reality around us. It may be regarded as an effort to locate, within the technological, all that we consider beautiful and sacred. If we are living in a simulation, then love, family, ritual and meaning are all products of the same technology that produces the simulation. Therefore, to claim the technological *as human*, and capable of producing human salvation and human love and human justice, we perform a digital enchantment of the world. As society marches towards a technological totality, the simulation hypothesis may be an attempt to read ourselves back into the fabric of reality. If the universe is information and matter is an illusion, then locating some agentive nature in us getting to this point may be part of healing that divide. If reality is a computer program, and we are a part of that reality, then we as humans are connected to each other and to Nature in some meaningful, tangible way. Perhaps there is a wisdom within the collective unconscious that understands that we cannot lose our mythological knowledge and remain healthy as a species. So, a new myth emerges, told in technological language, that keeps a certain kind of continuity intact. Like many myths before it, the simulation hypothesis places us in a reality in which time is a circle: the future is our past.

It may be worth mentioning that the simulation hypothesis became popular in the first few years of the recent millennium. Right before the year 2000 there was another archetypal structure that exploded in the technological imagination: Y2K. In this instance, the archetypal formation was an apocalyptic one, the direct counterpoint of a creation mythology. The word apocalypse comes from the Greek word *apokalypsis*, and refers to the revelation of that which has been hidden. The verb *kalypto* means "to cover or hide," and the preposition *apo* means "away." So together it means the lifting of the veil that exposes that which has been covered.[18] In a Jungian interpretation, apocalyptic archetypes indicate the end of an age. Much like how creation myths constellate as a new worldview emerges, apocalyptic myths may appear as a worldview declines. It is interesting to consider why the Y2K phenomenon may have preceded the simulation argument. It is as though we had to imagine a way that technology might cause the end of the world before we were able to see it as something that could generate the world's beginning.

A creation myth is a description of the process by which nothingness becomes something. Because it is unknowable, the event of the creation of the universe remains, both scientifically and mythically, under the influence of archetypal structures. We cannot know that which exists beyond our phenomenal reality, and the only thing we have immediate knowledge of is the content of "the psychic image reflected in consciousness."[19] As we become more embedded in our technologies, our psychic images of the human, the mind and the world are transforming. They will undoubtedly change many more times as we become more deeply interconnected. However, at this point *now*, where we find ourselves in a hybrid state of existence, on a bridge between the future and the past, we encounter the simulation hypothesis. The hypothesis itself may not be considered fully implementable, because of its own reliance on the future-real, but certainly in its openness and form we find a corresponding psychic reflection of our current human condition. There is something beautiful in the proposition that even as we progress towards a more technologically dominated reality, we are still weaving myths—although mythologies, and the realms in which they appear, are transforming in this same process.

The philosopher David Chalmers has said that we will never have any conclusive proof that we are not living in a computer simulation, because any proof we might come to discover could itself be simulated. As it stands, the inquiry around whether or not our world is a computer simulation falls in a long line of questions that remain without answers. In this very way, instead of telling us where we are going or where we came from, it orients us to where we are *right now*—caught somewhere between knowing and unknowing, matter and spirit, atoms and pixels.

The Myth of Matter

PART I

Everything we call real is made of things
that cannot be regarded as real.

—Niels Bohr

The book you're holding doesn't exist. The paper it is made of doesn't exist, nor does the ink on its pages. In fact, neither do you. Although it appears to our crude sense as real, the *matter* that makes up this book, you and everything else in the apparent universe *does not exist at all.* When reduced to the tiniest observable scale the atoms in the physical world reveal themselves to be constructed from a ghostly soup of *immateriality*. When traced to their absolute core, the very roots of matter transcend the material itself. The physical reality we inhabit arises from a nonphysical realm, which organizes itself into the spacetime dimension. The idea that our universe is composed of a concrete matter that is stable and real is an illusion. Matter is a myth.

Until the early 1900s, Western science thought that it had figured out a reliable picture of the physical world, save for a few small issues here and there. Before the advent of quantum mechanics, the field of physics was governed for nearly two hundred years by the classical laws described by Isaac Newton. Newton's universe consisted of stable objects in empty space. This rational view of nature was continuously reinforced because the laws of classical physics provide substantial, useful predictions about physical processes in the macroscopic world in which we live. But we know now that reality extends *far beyond the dimension of our everyday lives*, and in the first thirty years of the twentieth century, the Western understanding of what was *real* underwent tremendous transformation. From the rapid growth of technologies, which allowed us to engage with levels of nature far beyond our well-worn human scale, emerged a picture that obliterated the absolute validity of traditional rational materialism.

The subatomic theories and experiments of the twentieth century showed that matter and so-called empty space were not only intimately connected, but may in fact be the *same thing*. Quantum physics drew a new picture of reality: Not only was the division between subatomic particles and the so-called empty stage of space illusory, but so was the separation between an independent objective reality that exists on its own and an observer who witnesses such a reality. The view of the solid empirical world embraced by classical physics faded away into a mysterious, irrational reality-substance, where matter and nothingness, and the observed and the observer, were seen to be constitutionally interwoven, if not one and the same.

If these dizzying conclusions are correct, and the world of matter that we see as being discrete and discontinuous is an illusion, then still something must be responsible for establishing that illusion. The primordial dimension that lies beneath the subatomic world has been envisioned and analogized in many different ways by scientific thinkers in the last two centuries—as has the relationship between this dimension and our perceptible reality. This dimension has also been imagined in many ways outside the scientific paradigm. Mythic and religious structures have, for millennia, described a realm beyond the actuality of spacetime, from which the three-dimensional world arises. In this way, science and religion—and by extension mythology—share a goal: the comprehension of reality.

The questions at the heart of these considerations are these: How does matter emerge into the universe, and from where? How does the elementary grammar of our reality arrange itself into the language of the physical world? The emergence of something from a perceived nothingness can be considered from many different theoretical perspectives. In trying to envision that which is incomprehensible, the human mind has produced a kaleidoscopic array of structures, some of which appear to be shared across different worldviews and contrasting disciplines. This can be seen quite strikingly when quantum theory is placed in conversation with various Eastern religious, philosophical or mythological schools of thought (namely Hinduism, Buddhism and Taoism). The possible correlations between various forms of mysticism and the view of reality ushered in by quantum mechanics reveal, at minimum, a connectedness in the ways humans have attempted to think about the unthinkable, with metaphor and analogy playing critical roles in both schools of thought.

So, what exactly is the whatness of matter? This essay will attempt to address this question in two parts. This first part will look at the ways Western science has attempted to answer this inquiry, from both mechanical and scientifically allegorical perspectives. The second part will look at matter from a different angle, via its relationship to mind as theorized by Carl Jung. As a whole, this essay asserts that matter is a myth in two ways. The first, which will be addressed in the following first section, is that physical matter is a myth in the sense that it does not exist. It is a false formulation based on the inability of our coarse senses to apprehend the true nature of reality. The second is that physical matter is a myth in the sense that it is not separate from our own consciousness, and therefore does not have an intrinsic being-in-itself.

The topic is vast, potentially infinite, and this essay will in no way exhaust the endless mysteries about matter and reality discussed in quantum theory or mythological paradigms—and will certainly ask more questions than they answer. Readers familiar with the subjects may feel as though important things have been left out, while others may wish that certain subjects were explored further. As a layperson, what I have tried to do is provide an overview of these topics, in which readers will hopefully discover a portal or two to their own curiosity and understanding. In service of this goal, we will begin with a short history of the atom...

The discipline of physics gets its name from the Greek word *physis*, which translates as Nature. Although physics is one particular branch of science, it has important implications in the other sciences as well. Biology is the study of organic cells, which in part is also the study of the molecules that make up those cells. Chemistry is the study of molecules, which in part is also the study of the atoms that make up those molecules. And physics is the study of atoms. Atoms are the elemental unit of the physical world and are almost inconceivably small: It would take about 250 million of them arranged in a line to reach a single inch. A small drop of water consists of more than 20 billion molecules, each of which is made up of three atoms. And in order to see the atoms in a soccer ball, you would have to enlarge the soccer ball to roughly the size of the Earth.

Matter can be defined as any substance that takes up space with its volume and possesses some kind of mass, and atoms are the building blocks of matter. A definition of an atom might be: an individual, self-contained entity that has the same number of protons as electrons. Another definition might be that an atom is the minimum quantity of a chemical element that can exist. Atoms consist of several parts; a positively charged central core called a nucleus, which is comprised of *sub*atomic particles called protons and neutrons, and which in turn is encompassed in a cloud of electrons. Protons carry a positive electric charge, electrons carry a negative electric charge and neutrons have no electric charge. The iconic image of the atom, which shows the nucleus surrounded by electrons that travel on stable ringed paths, like planets orbiting the sun, is beautiful but incorrect.

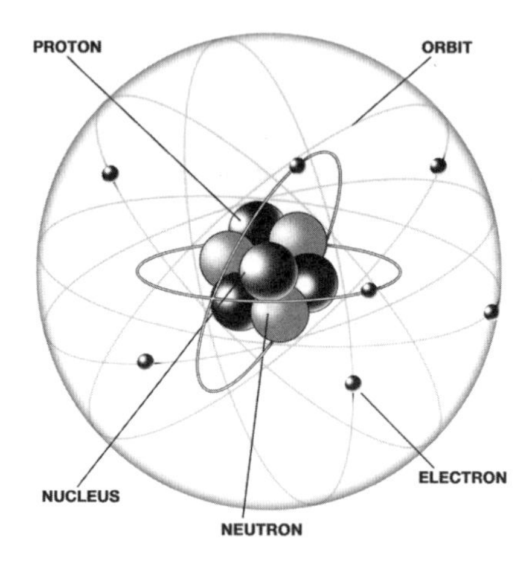

Instead, atoms look more like fuzzy little orbs, with a nucleus at the center and electrons surrounding the nucleus in a foggy spherical puff.

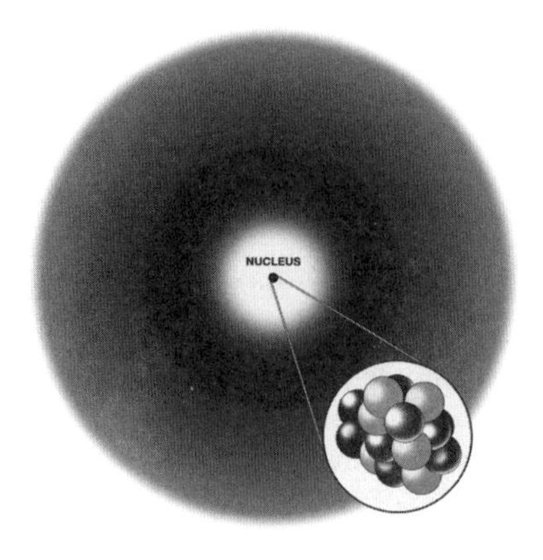

This electron puff is somewhat like a planet's atmosphere; it does not have a defined outer limit and instead dwindles out gradually. Every kind of atom, except the helium atom and the hydrogen atom, has two or more of these electron clouds arranged concentrically around the nucleus. When atoms unite to form molecules, their electron puffs overlap and create different shapes. It is important to note that these clouds of electrons are fuzzy not because of the speed at which the electrons themselves move, but instead because of something *strange and incredible that happens at the subatomic scale of reality*—but we'll get to that shortly.

The original concept of the atom is often attributed to the ancient Greek philosopher Democritus, who lived from 460–370 BCE (although similar ideas were being developed in India by Buddhist, Jain and Hindu philosophers at that same time). Democritus, like many other Greek philosophers, was concerned with uncovering the foundational components of matter and the underlying order of nature. He proposed that all matter could be divided into smaller and smaller pieces until a certain point where matter could not be divided any further. At that scale, according to Democritus, matter was made up of tiny, imperceptible particles he called *atomos*, a Greek word that translates as "that which cannot be cut." Democritus suggested that solid matter consisted of weighty, tightly packed atoms, whereas gases consisted of much lighter atoms with a lot of nothingness in between them. In his theory, this concept of nothingness—empty space or the "void"—played a crucial part. To Democritus, these atomos particles arranged themselves with relation to the void or nothingness in which they existed. Democritus's views were not very popular. His was a materialist view of the world, and that did not sit well with the vast majority of thinkers at the time who leaned, at least in some capacity, toward a religious and/or metaphysical view of nature. Another reason why Democritus's ideas were not embraced was because of their reliance on the concept of nothingness or the void. Aristotle, who lived about one hundred years after Democritus, considered Democritus's atomic ideas to be particularly foolish. Aristotle believed that the matter which made up the physical world was, in essence, continuous and endlessly extended. According to Aristotle, matter itself had no beginning or end, and was infinitely divisible. Within his belief system, there was no room for the concept of nothingness or the void—according to Aristotle there was no such thing as empty space. Even if space looked empty, it was actually full of gas or air, and therefore could not be considered a true void.

Aristotle's ideas about nature prevailed in the West until the seventeenth century when a scientific perspective began to outweigh a philosophical one. By the mid-seventeenth century, the view that matter was divisible and made up of particles started to reappear, as did the concept of the void. In 1659, the scientist Robert Boyle created a vacuum pump, a device that removed the air from a glass dome in which certain experiments could be performed. This invention reignited debates about whether or not there could be a space in which no matter existed. Boyle's other, and arguably larger, contribution to the development of atomic theory was in redefining the concept of an element, which helped to kill the pre-Aristotelian notion that there were only four elements—earth, air, fire and water—and that they were the basic building blocks of physical nature. According to Boyle, an element was any kind of substance that cannot be broken down into simpler substances. In this way of thinking, water could not be an element, because it can be broken down into oxygen and hydrogen. Boyle championed a kind of atomism called *corpuscularianism*, and within this framework, Boyle defined elements as "primitive," "simple," and "unmingled bodies," which are not made up of other kinds of bodies, but are in fact the "ingredients of all those so call'd."[1]

Corpuscularianism was also adopted by Isaac Newton. Like Boyle, Newton considered the world to be mechanical in nature. For Newton, the universe consisted of stable objects in empty space. He considered matter to be passive and made up of "solid, massy, impenetrable, movable particles."[2] According to Newton's laws, a body in motion will remain in motion and a body at rest will remain at rest. In order for a body to move at all, it must be acted upon by some external force, but matter itself was inert by its own nature. These formulations, along with Newton's theory of gravity and laws of motion, ushered in the reign of "classical physics," and helped resolve the seventeeth-century Western dispute over whether science or religion was the ultimate ground of truth, with science prevailing.

Around 1800, a chemist named John Dalton was studying the pressure of gases and the properties of chemical compounds (substances composed of two or more separate elements). Drawing directly from Democritus, Dalton believed that all matter consisted of tiny particles. His work with gases convinced him that they were also made up of tiny particles in constant motion. Through his experiments, Dalton concluded that compounds always consist of whole-number ratios of atoms (for example: water must always be decomposed into two atoms of hydrogen and one of oxygen—the familiar H_2O and never a fraction of an atom like $H_2O.35$). He theorized that this was due to the fact that atoms, which made up compounds, could not be divided any further. There could never be anything but single, whole atoms. He therefore hypothesized that atoms themselves were indestructible and unchanging, and that the atom of any particular element must have the same mass as all the other atoms that make up that same element (this idea eventually led to the concept of "atomic weight"). Dalton further proposed that atoms belonging to different elements would somehow be fundamentally different from each other, and that compounds were formed when the unique atoms belonging to different elements combined. Although some of Dalton's hypotheses would be proven untrue, many were either correct or laid the foundation for future successful theories. What is particularly remarkable about Dalton and his ideas is that he built his theories entirely based on deductions he made from macroscopic—not microscopic—experiments.

In the late 1800s, a physicist named J.J. Thomson discovered the electron. His discovery would demonstrate that the atom was, in fact, divisible. While performing experiments using cathode rays (glowing rays that are produced when voltage is applied to electrical conductors in a glass tube), he discovered that the rays were repelled by negatively charged metal plates but attracted by positively charged ones. Thomson realized that this meant that the rays themselves must then be negatively charged! By performing a series of experiments based on this discovery, he concluded that the mass of the negative charge was over one thousand times lighter than the hydrogen atom, which was the lightest matter that had been discovered at that point. Thomson called the impossibly light, negatively charged particles that made up the ray *electrons*. By using different kinds of metal plates, he was also able to deduce that these electrons could be found in many different types of atoms, therefore demonstrating that atoms themselves were made up of smaller *subatomic* particles. From this, Thomson came up with his own atomic model called "the plum pudding model," which proposed that atoms were spheres that were positively charged throughout, with electrons (which he called corpuscles, keeping with the theory of corpuscularianism) dotted randomly within them.

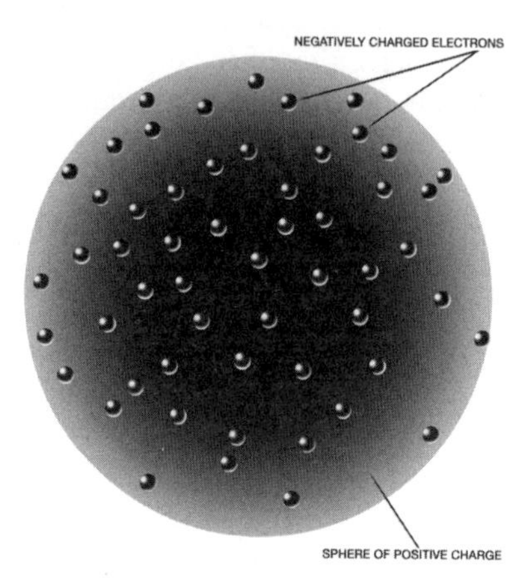

Thomson's model would soon be disproven by a student of his, Ernest Rutherford. By performing experiments in which he fired a stream of positively charged alpha particles (composite particles consisting of two protons and two neutrons) from a radioactive source at gold foil, Rutherford was able to discern that the positive charge of an atom was not spread out, randomly or otherwise, within a sphere. Instead, the positive charge, and most of the mass, of an atom was contained at the center. He called this center the nucleus. So, Rutherford proposed a model of the atom called "the planetary model," which saw the atom as being similar to a solar system, with electrons orbiting the nucleus in a broad and steady trajectory. However, the theory did not explain what kept the electrons in orbit and prevented them from collapsing into the nucleus. According to the laws of classical physics, the electrons envisioned in this model of the atom ought to gradually lose their energy, moving toward the center of the atom in ever-weakening circles until they eventually slam into the nucleus. But this was not happening. Although it would take some time for that mystery to be solved, something else extraordinary was discovered in Rutherford's experiment, something that set the foundation for the entire upheaval of classical physics: *Most of an atom was actually just empty space.*

This was the first indication that matter was truly bizarre. Rutherford's experiment revealed that the atom was not the solid and massy particle it was thought to be since the days of ancient Greece. Instead, atoms were objects comprised mostly of regions of empty space! With this, the notion that the fundamental unit of matter, the atom, was a solid object—which was the cornerstone of scientific thinking up until that point—was eradicated. As wild as this discovery was, it was only the first of many that would turn the world of the physical sciences completely upside down.

Concurrent to these developments, there was another scientific marvel bubbling to the surface: quantum theory. By the end of the nineteenth century, it was recognized that matter came in two forms. The first was particles, as we have so far discussed, and the second form was *waves*. At the end of the seventeenth century, scientists like Newton had believed that light was a kind of "shower" of tiny corpuscles. But this all changed in the beginning of the nineteenth century with a scientist named Thomas Young. Young's double-slit experiment demonstrated that light displays the properties of a wave, and then when waves of light combine with one another, they create *interference patterns*. The experiment goes like this:

Light from a light source is shone onto a screen with two slits in it, and behind the screen with two slits there is a second screen. As the light passes through the two slits, two secondary light waves emerge and spread out, overlapping and crossing each other. As the two secondary waves meet, they create a cross-hatching pattern on their way to the final screen, on which an interference pattern can be seen.

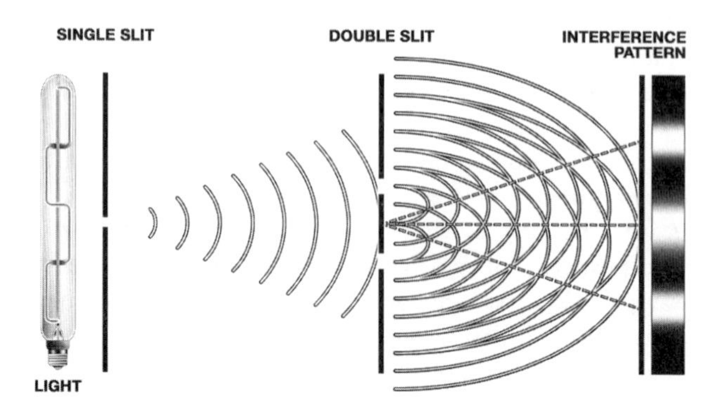

These interference patterns are the same ones you would see if you were to drop two rocks into a still pond and watch the ripples pass through each other. On the back screen, a sequence of light and dark stripes appears. The lighter stripes, which are bright bands of light itself, are caused by *constructive interference*. Constructive interference happens when the crests (or highest points) of the two secondary light waves overlap and merge with one another. This creates a more intense light, which is then seen brightly on the back screen. However, where the crest of one of the secondary waves meets the trough (or lowest point) of the other, they cancel each other out, and this creates the dark bands on the back screen. Some light remains between these two extremes, which results in a kind of moderate blending around the strong bright bands. These kinds of interference patterns could *only* occur if light behaves like a wave. However, although Young's experiment proved that light behaved like a wave, it did not tell us what light was a wave *of*.

The scientist James Maxwell answered this question in 1865. Maxwell first showed that magnetic fields and electrical fields were essentially two aspects of the same phenomenon, now known as electromagnetism. He then demonstrated that light was an electromagnetic wave, generated by oscillating electric and magnetic fields. As such, light could be described in terms of *wavelength*, which is the distance between two crests of a specific wave, *frequency*, which is how fast a specific wave goes up and down, or vibrates, and *amplitude*, which is the measurement of how far the wave rises above and how far it falls below its origin point.

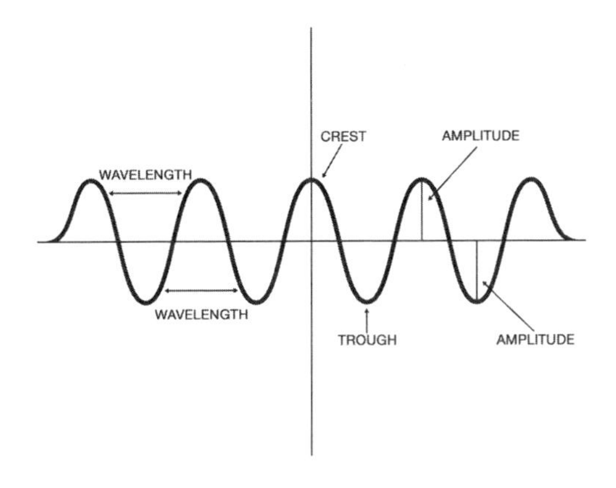

However, Maxwell's picture of light as *only* a wave crumbled when Albert Einstein began researching the "photoelectric effect." The photoelectric effect, first observed in the late nineteenth century, is a phenomenon in which the electrons on a metal surface are ejected when a light is shone onto the metal surface. When the electrons are discharged from the metal, they become known as "photoelectrons." Because light was thought to be just a wave, Maxwell's theory led to the conclusion that the *intensity* of the light (how much energy the wave carries) was responsible for the process of ejecting the electrons from the metal surface, the act of which generates photoelectrons. However, it was later discovered that it was actually the *frequency* of the light wave that determined whether or not the electrons on the metal surface would be dislodged. If the light had a high frequency, meaning that the light wave went up and down rapidly, resulting in a high vibration, it would produce photoelectrons, even if the wave had low intensity (energy). However, if the frequency of a light wave was low, the process would not occur, even if the intensity of the light itself was high.

In 1905, Albert Einstein explained this phenomenon by hypothesizing that light was in fact composed of particles, each of which possessed a specific amount of energy. The intensity of any given light was related to the number of particles in a specific light beam, and the frequency of the light corresponded to how much energy each of the particles carried. So, to produce the photoelectric effect, the amount of energy in the particles that make up the light wave (the frequency) had to be high in order to knock the electrons off of the metal surface. If the individual particles themselves do not possess enough energy then the electrons on the metal surface will not be moved, no matter how many particles there are (the intensity).

Einstein referred to these indivisible packets of energy that made up light as *quanta*.[i] This built off the work of the physicist Max Planck. Planck had originally introduced the concept that would become quanta through his experiments with "black body objects." Black body objects are theoretical objects that absorb all the radiation they receive and do not reflect any light or allow any light to pass through them, but which do emit thermal radiation energy at temperatures above absolute zero. Planck calculated that the heat radiating off of black body objects was not emitted as a continuous wave. Instead, the thermal radiation from black body objects appeared to be collections of tiny, distinct "packets" of energy. Planck called these packets quanta, which he derived from the Latin word quantum, which means "sum" or "amount." Planck employed "quanta" to describe the *very minimum of any quantity that can exist*. Planck's mathematical experiments were the first to demonstrate that, in certain situations, energy can display the properties of physical matter.

Although Einstein eventually won the Nobel Prize for the theory in 1925, his light-quantum explanation of the photoelectric effect was considered reckless by other scientists when he first proposed it in 1905. What made the explanation so astonishing was that, in demonstrating that light was both a wave *and* made up of particles, Einstein showed that matter and energy were, in essence, interchangeable. Furthermore, although particles and waves in ordinary experience are mutually exclusive kinds of substances, light was both at the same time—or rather, humans are able to interpret light as both a particle and a wave at any given moment, depending on the interactions observed.

Now back to the atom: although he disagreed with Einstein's light-quantum hypothesis until the 1920s, the physicist Niels Bohr did use quantum concepts to build a new atomic model in 1913. According to the laws of classical physics, the electrons in Rutherford's

---

i These individual packets of light quanta would eventually be called "photons."

solar system model should eventually lose energy and spiral down in ever-decreasing loops, ultimately crashing into the nucleus. However, this was not observed, so something was off in Rutherford's configuration. Bohr looked to Planck's quantum theory, which said that only a certain amount of energy was *allowed* in any given system at any given time (and that systems were composed of these discrete energy packets called "quanta"). Bohr proposed that the electrons in an atom revolve around the nucleus in fixed, circular paths determined by their size and energy. The lower the energy of the electron, the smaller the orbit and the closer to the nucleus. Because the orbit itself would count as a system, once the energy level of one orbit was filled up to maximum, a new energy level (orbit) would begin one level above it, like the rungs of a ladder. Because of this, Bohr suggested that electrons do indeed have stable paths and called these "stationary orbits."

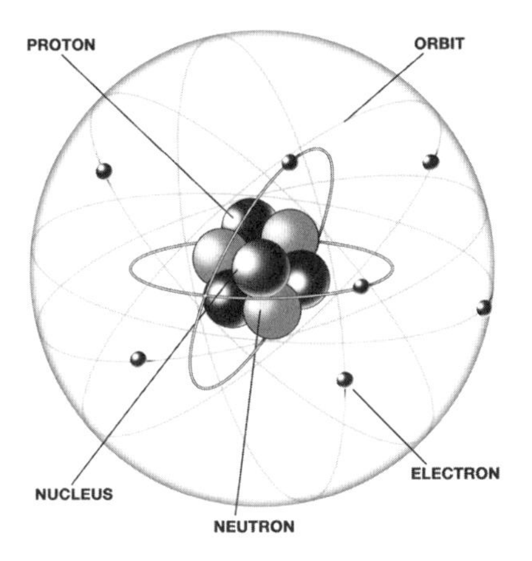

When electrons increase in energy, which happens when they *absorb* incoming electromagnetic radiation, they jump up to a larger/higher orbit. When they decrease in energy, which happens when the electrons *emit* electromagnetic radiation, they jump down. This meant that the energy inside electrons were quantized as well, and the specific amount of energy in each electron determined the size of their orbit. Electrons could only move between these specific energy levels—symbolized by the orbital path—by either gaining or losing energy. This movement or jump between orbital levels is where we get the concept of a "quantum leap."

Then in 1924, the physicist Louis-Victor de Broglie made the radical suggestion that perhaps the wave-particle duality that Einstein discovered in light might also apply to objects that were considered to be only particles—specifically electrons. If light had momentum, energy and a wavelength, and matter had momentum and energy—could it be possible that matter also had a wavelength as well? In 1926, Erwin Schrödinger created an atomic model that proposed just that. Schrödinger hypothesized that the activity of electrons within atoms could perhaps be explained by viewing them mathematically as *waves*. In Bohr's model of the atom, electrons are treated solely as particles, which move in fixed orbits around the nucleus. In Schrödinger's model, however, electrons are treated as waves, which occupy three-dimensional space throughout the atom. By their very nature, waves do not have an exact location; waves constantly move. If an electron behaves like a wave, it too cannot be said to have an exact location. Instead of existing only on the stable orbital paths, as in Bohr's model, an electron can potentially be found anywhere in the atom in Schrödinger's model. Because an electron can potentially be found anywhere in the atom, Schrödinger's model—also known as the "quantum mechanical model"—does not try to determine the electron's precise location at any given moment. Instead, this model describes where the electron is *most likely to be found*. This model is still considered to be the most accurate model of the atom to date.

At first Schrödinger thought that the electron wave meant that the electron was somehow smeared throughout the atom. After all, if there is a wave, then there must be something doing the waving. In an ocean wave it is a collection of water molecules that make up the wave; in a sound wave, it is the pressure from the vibration moving through the gases that make up the air. So, Schrödinger, perhaps understandably, first believed that it was the electron itself doing the waving: that somehow the electron was extended in the atom's three-dimensional space and behaving in a wavelike manner. However, the physicist and mathematician Max Born soon showed that what made up the wave was in fact much, much stranger.

Born was able to recognize and prove that Schrödinger's wave was actually a *probability* wave. What this means is that *we can never actually know where an electron is going to be.* We can only predict the probability of where the electron will be, and this probability is distributed throughout the entirety of an atom like a *cloud.*

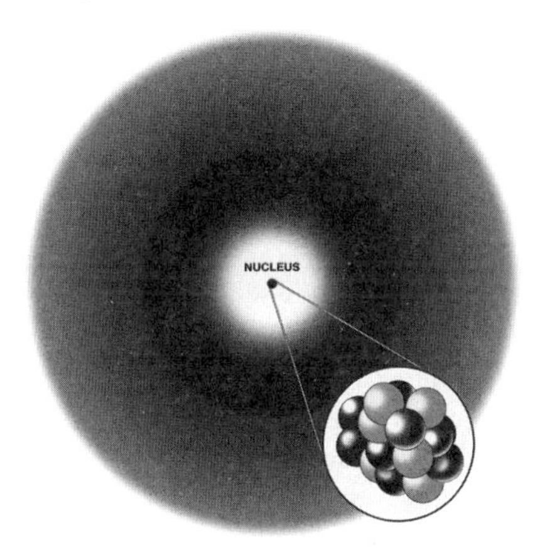

These clouds describe the likelihood that the electron will be found in any certain position. In areas of the atom where the electron is mostly likely to be found, the probability cloud is dense. In areas where the electron is less likely to be found, the cloud is less dense. These areas, where the electron is most likely to be found, correspond to the orbital paths described by Bohr, but the clouds are not necessarily spherical. Some are dumbbell shaped, some look like four-leaf clovers.

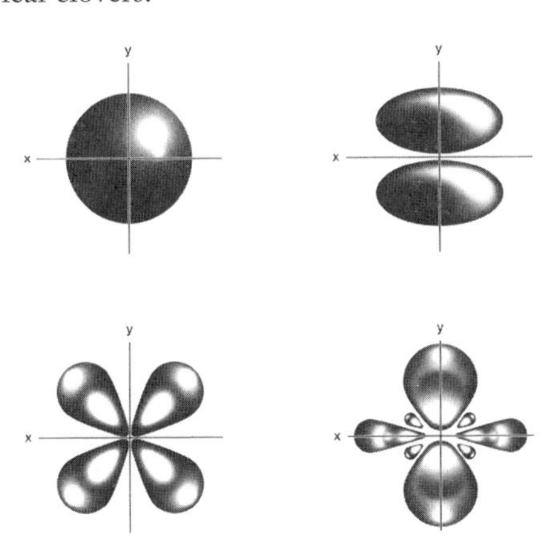

Although electrons do sometimes appear to be localized in a particular part of an atom, there is no guarantee that we will actually find them there. Because of this, electrons *cannot be said to have precise physical locations*. One can never ask where an electron is at any given moment, one can only ask, "If I were to look at a specific area of an atom, what is the likelihood of finding an electron there?" This means that the electron cannot be thought of as a *fixed object*. Instead, an electron is more like a *collection of possibilities*.

Although hard to conceptualize, this paradoxical truth says something fundamental about the nature of the electron, and therefore the fundamental nature of matter itself. Particles and waves are inherently different. A particle is a finite thing and it exists in a definite position. If a particle collides with another particle, it will bounce off and assume a different definite position. A wave, on the other hand, is something in *motion*. Unlike a particle, which holds all of its energy in a single point, a wave is spread out and its energy is distributed over space via its motion. If a wave collides with another wave, the two waves will pass through each other and interfere, just like the light waves in Young's double-slit experiment.

However, electrons can behave as both particles *and* waves. It is a concept that defies common experience and exceeds the scope of human comprehension. We naturally ask, "Well, which is it? Is an electron a particle or a wave?" The answer is *neither*. We actually do not know exactly what electrons are, and there are a lot of contesting theories. The only thing we can say with certainty about the electron is that it is a fundamental aspect of nature, a subatomic particle that possesses *dual-nature*, and can be expressed and observed as both a particle and a wave. When measuring an electron's position, it is observed to be a particle. When measuring an electron's momentum, it is observed to act as a wave.

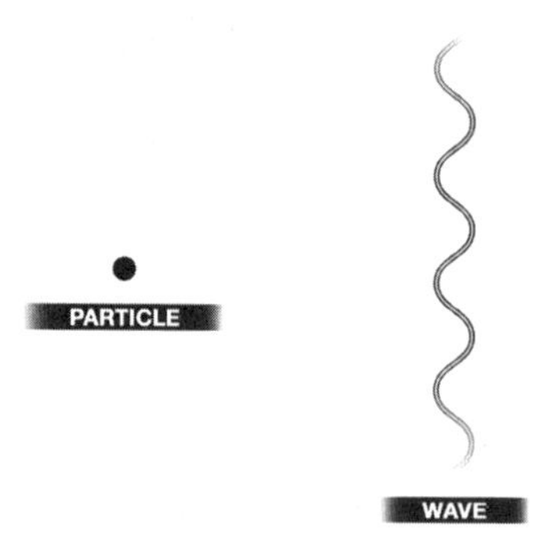

This concept is called the *wave-particle duality* or *complementarity*. It is an incredibly difficult idea to hold but it is the closest description of the actual nature of matter that we can get to in language, and is essential for all quantum thinking.

Complementarity is the notion that the fundamental units of matter possess complementary qualities that cannot be observed or measured simultaneously. This is Werner Heisenberg's uncertainty principle: we can never know both a particle's position and its speed/momentum at the same time. This can be imagined by returning to the analogy of ripples on a pond's surface. If we wanted to measure the momentum or speed of a ripple, we could observe and measure the movement of its different peaks and troughs. The more peaks and troughs we see and measure, the more accurately we can deduce the speed of the ripple's wave. However, if we were interested in knowing the precise position of one specific peak of the wave, we would zero in and measure one tiny portion or point of the wave. But, in doing that, we would lose information about the wave's momentum/speed. The two properties—momentum and position—can never be observed at the same time.

You can get a rough idea if you look at this image:

Viewed one way, you see two faces in darkened silhouette, viewed another way, a chalice or goblet—but it would be hard to claim that you can see both perspectives at the exact same time.

The striking weirdness of the wave/particle duality is also demonstrated beautifully in another kind of double-slit experiment. If we perform a double-slit experiment similar to Young's, but this time with electrons being fired at the two slits instead of a beam of light, we will see that the electrons behave in ways that are hard to comprehend.

Imagine an experiment where an electron gun, which shoots beams of identical electrons, is positioned in front of the first screen with two slits in it. Then, behind that first screen is a second screen capable of recording where an electron has hit it by registering the electron's position with a little bright dot.

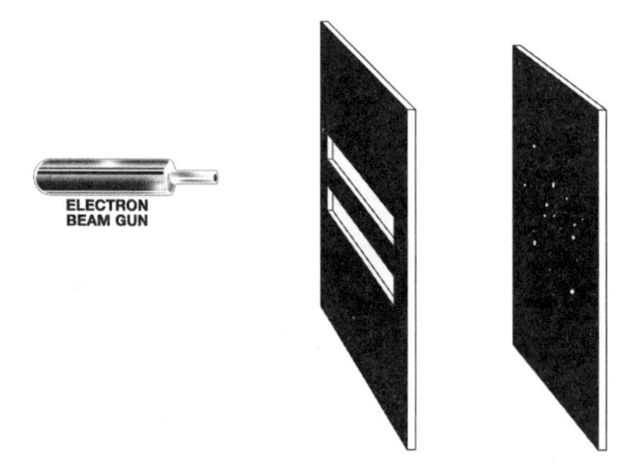

The electrons leave the gun as localized particles and are recorded on the back screen as localized particles, but what happens in between exemplifies the incredible bizarreness of subatomic particles.

Because we observe the electrons leaving the gun as particles, we might logically assume that as the particles hit the first screen, some of them will go through the top slit and some through the bottom. As particles, we might logically expect the electrons to behave like grains of salt, forming two piles behind or underneath the two slits.

However, this is not what we see. Instead, we see a pattern on the back screen that is remarkably similar to the patterns recorded in Young's original experiment with light. Bright stripes start to appear where the electrons have hit the back screen, with the brightest stripe appearing in the middle between the two slits and not directly behind them, which one would expect if it was assumed that electrons were merely particles. This is an interference pattern, which can only be created through the interaction of waves.

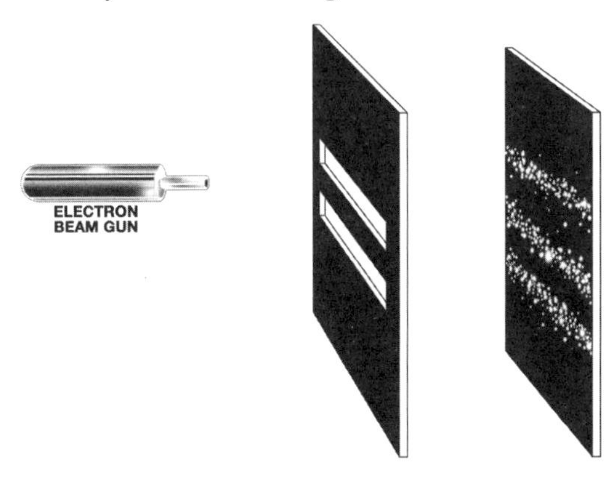

When the light wave from Young's experiment hit the two slits on the first screen, it was diffracted and split in two. The crests and troughs of those two waves then interfered with each other on their way to the final screen, causing the interference patterns on the back screen to appear. This is exactly the pattern that is seen with the beam of electrons, even though they leave the gun as particles.

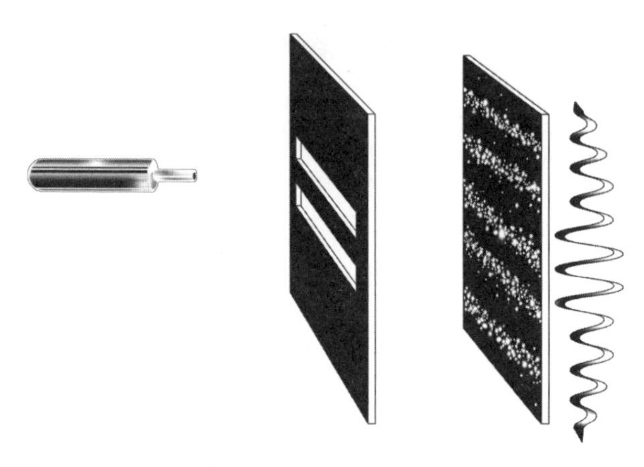

What is even weirder is that this pattern will be the same even if we send the particles *one by one.* Somehow, even though they are isolated, the electrons still behave *collectively* as a wave. One by one they pass through the slits, cross and interact with each other as two waves would, and hit the back screen as a particle, slowly building up a pattern that is directly correlated with the interference between two waves. But what could the electrons possibly have interacted or interfered with if they were fired one at a time? A wave is the result of "collective behavior," and yet even as the particles pass through the slits one by one, they demonstrate this wavelike behavior. Furthermore, we can never predict exactly where a single particle will strike, we can only formulate a pattern of distribution that predicts where the hits might occur. But because the electrons are identical, there really isn't any reason why one particle should land in one place and another in another.

Even more bizarre, this wavelike behavior only occurs when there are *two* slits open on the first screen. If one slit is closed off and only one slit open, all of the particles will go through the open slit and form a single band instead of an interference pattern. However, this does not mean that the electron somehow splits in two when it goes through both slits. If that were the case, we would see half an electron go through each slit—but that does not happen. Detectors placed right at the two slits record that only one electron goes through a slit at a time, *always*. It never splits in two. Somehow, it appears that the electrons are aware of when there is one slit open and when there is two. We have no idea how or why this occurs. The electrons are particles when they leave the gun, and particles when they arrive at the backscreen, but something odd happens in between. Even if they are fired one at a time, the electrons exhibit the behavior of a wave that spreads out, splits in two and then interferes with itself. Moreover, the same results have been seen with experiments using other atomic and subatomic particles as well, such as photons and buckyballs, which are groups of sixty carbon atoms in the shape of a soccer ball and are much, much larger than an electron.

Just what then are these waves made of? As hard as it may be to understand or accept, they are waves of *probability*. Probability waves are the foundation for the physical world, although they are not themselves a physical entity. Nor are they a metaphor. The wave of the particle/wave duality is made up of probabilities—in the very realest sense that this can be imagined. The thing that is doing the waving is all the possible places the electrons could be. Yet when these waves of uncertainty get to the back screen, the electron must decide where to land as a particle. But probabilities are numbers, ratios of numbers in fact; they are mathematical descriptions about how likely something is to occur. Probability waves are intangible, they have no mass or energy, and yet the entire physical universe and the definiteness of matter is determined by the interference of such waves. Physical objects are localized, but the very nature of the particles that comprise any physical body *cannot be said to be localized*. Atoms are not tiny beads of solid matter, but instead are standing waveforms made up of mostly empty space, in which electrons can only be described as *abstract entities*.

And yet these abstract entities are responsible for the *solidity* of the solid physical world! Before an electron is measured, it exists in the atom as a probability wave. Once an electron interacts with the physical world—as we measure it—its *wave function*, which is the mathematical description that defines the electron's behavior as a wave, "collapses." This means that when the position of an electron is measured, its wave function instantly becomes zero in every position we know the electron isn't. If a probability wave were a physical entity, this collapse could not be instantaneous. But experiments show that the collapse is instantaneous. If we understand and accept then that the wave is abstract, irrational and purely mathematical—and that the properties of subatomic particles transcend materialism as we know it—then this instantaneous collapse poses no problem. An electron's wave function describes this behavior with incredible accuracy, but does not explain it.

If this sounds crazy or counterintuitive—that's because it is, at least according to our concepts of reality derived from everyday experience. The wave-particle duality of the elementary units of matter defies the ordinary understanding of matter as a definitive thing. The very basic structures of the material world are themselves indefinite. Probability is an immaterial, informational, nonphysical thing, yet probability is responsible for the physical components of an atom. We have no choice but to conclude that the order of the physical universe is based on nonphysical phenomena. The foundation of reality as we know it is nonmaterial.

As hard to accept as this all might be, we know the laws of quantum mechanics to be correct. Since their inception, quantum laws and concepts have been used to make predictions in all different kinds of experiments, and time after time, the tenets of quantum mechanics have proven true. In fact, there has never been any disagreement or conflict between quantum laws and any experiment that has ever been done! Much, if not all of the information technology that rules our modern existence—like phones, computers, digital cameras—would not exist without these quantum concepts. These technologies are controlled by microscopic switches that direct and guide the currents of tiny electrons, and the equations that inform these inventions come directly from quantum mechanics. However, even with experimental and practical confirmation of a quantum world, physicists, mathematicians and philosophers still argue about the fundamental nature of quantum theory and the theory's implications for our understanding of reality.

Even the scientists involved in the development of the theory had a hard time accepting it. Heisenberg and Bohr would have desperate, late-night discussions trying to wrap their minds around the absurdity of what their experiments revealed. Heisenberg's solution was to recommend that the paradoxical (or complementary) nature of subatomic particles simply be accepted, as well as the reality that subatomic entities can only be described in indefinite terms. Einstein, as brilliant as he was, had particular trouble with this. He couldn't believe that subatomic particles and atoms, which make up all the physical matter in the universe, were simply governed by probability and not by certainty. Einstein could not accept that nature could be so fundamentally random and famously remarked "God does not play dice with the universe." To which Bohr, who considered complementarity to be the only actual, objective description of nature, responded, "Stop telling God what to do." Bohr was able to tolerate the notion that the foundation of reality was blurred and fuzzy. Einstein could not.

Another point of contention for Einstein was the notion of *quantum indeterminacy* and what happens when the wave function collapses. Quantum indeterminacy is the idea that a subatomic particle can exist in one of multiple states, and that you can't know what state it is in until it is observed. We saw that an electron cannot have a definite momentum and a definite position at the same time. However, this lack of definition goes beyond our inability to know both properties. Instead, it means that the electron *itself* does not possess both properties at once. The act of observation somehow forces the subatomic entity to decide what state, and what location, to be in. Before subatomic entities are observed and measured, they exist in what is called a *superposition state*, which means that a system has two (or more) different states that could possibly define it. The act of observation somehow causes the wave function to collapse into only one state. Bohr and Heisenberg accepted that observation played a critical role in determining reality, and that quantum theory describes our *knowledge* of reality, not reality itself. Einstein, however, could not accept the premise that the reality of the entire universe depends on whether or not we are looking at it.

Einstein also took issue with a concept that, along with the wave function and superposition principle, lies at the heart of quantum physics—*quantum entanglement*. Simply put, two particles become entangled if they interact physically in such a way that their fundamental properties become inextricably linked. Even if these particles are thereafter placed very far apart (say one on Earth and one on Neptune), they will remain mysteriously connected and share a quantized state or condition (such as spin, a measurable property of angular momentum, which all subatomic particles have). No matter how distant the two particles are, if you measure a property of one particle such as spin, you will instantaneously affect the other. This kind of correlation is not something that can occur by chance; the particles are intrinsically related. What makes this so strange is that there is nothing tangible connecting the two, no force or material, that we can observe thus far, tying them together. It seems that, somehow, the entangled particles know something about the other and as soon as one is acted upon (say by measurement), the other is affected in an equal but opposite way.

Bohr accepted entanglement as a true description of reality. He believed that the act of measurement brought the first particle to a definite state, and that once the first particle was defined, the second entangled particle would instantaneously be defined as well.

However, Einstein had a real problem with this phenomenon, which he famously called "spooky action at a distance." The problem was that this instantaneous connection and response implied that somehow the two particles were sending signals to each other faster than the speed of light. However, Einstein had already proven through his theories of general and special relativity that nothing could possibly move faster than the speed of light. Because of this, he could not believe that the act of measuring one entangled particle somehow determined the states of both, and instead believed that each possessed a definite set of conditions before they were even measured. To Einstein, the act of measuring one particle of an entangled pair was similar to splitting a pair of gloves up into boxes and sending them to two different people on opposite sides of the world. At every moment, one box would always contain the left glove and one box would always contain the right glove, but you would not know which glove you were getting until you received the box and opened it. Once you have the box in front of you and look at what is inside, you can then immediately deduce what is in the other box. If you receive the lefthand glove, you know right away the other box contains the righthand glove. For Einstein, it was not the act of measurement that determined the state of one particle, and therefore the state of the other. Instead, the two gloves were always in the states they are eventually found in, we just hadn't looked yet so we could not know. This led Einstein to believe that quantum mechanics was in fact an "incomplete" theory, and that some new theory would eventually explain its unusualness. Einstein died in 1955, still believing that the theory of quantum mechanics was incomplete.

In the two decades that followed Einstein's death, quantum entanglement was experimentally proven to be true. In 1964 a physicist named John Bell published a paper that presented a mathematical way of figuring out whether or not particles could really be entangled. Then in the 1980s, a physicist named Alain Aspect created an experiment based on Bell's ideas, which proved that entanglement was physically real, not just a mathematical hypothesis, and later experiments provided additional confirmation. Einstein was wrong—quantum particles can be unbreakably linked throughout space.

However, because it remains true that nothing can travel faster than the speed of light, some physicists have suggested that the instantaneous relationship between entangled particles following measurement is not due to some mysterious super-fast signal, but is instead the result of an even more dizzying concept called *nonlocality*. At its core, nonlocality means that there are certain instances in which *location* itself does not exist. If this sounds crazy, that's because it is. To this day, many physicists still regard entanglement as the most problematic aspect of quantum physics.

Quantum mechanics is a set of mathematical rules, which were created by early twentieth century physicists who were studying the atom. As they did, they discovered that at reality's tiniest scale, the laws of classical physics break down. While the formulas of quantum mechanics provide reliable experimental predictions about the properties and behaviors of subatomic particles, the underlying meaning of the theory, and what it says about the true nature of reality, remains disputed. Because of this, there are many different *interpretations* of quantum mechanics, each of which seek to explain what the hell is going on. Some interpretations try to make adjustments so that quantum mechanics can be more fully understood within the three-dimensional physical paradigm, while others argue that it is not actually necessary to understand it fully at all. However, all of the interpretations have something to say about quantum indeterminacy and the collapse of the wave function.

The most widely known interpretation is the Copenhagen interpretation, named after the city in which Heisenberg and Bohr did most of their original work. The Copenhagen interpretation is an observer-dependent theory. It says that a phenomenon cannot really be considered a phenomenon until it is observed. Thus, the world becomes defined only through the act of measurement. According to this interpretation, the subatomic world appears probabilistic not due to some kind of incomplete knowledge on the scientist's part; instead that is the true reality of matter and energy at the subatomic scale. The Copenhagen interpretation says that the quantum world is inherently unknowable and that quantum mechanics can never describe exactly what happens at the microphysical level. Instead, it describes all that *we can know* about the world at that level. While we can theorize about how the world works, there will always be a level of reality we do not understand. However, this fuzziness does not make the predictive power of quantum mechanics any less correct or effective.

Another interpretation of quantum mechanics, and one that is perhaps even weirder, is the many-worlds interpretation. Conceived of by Hugh Everett in 1955, the many-worlds interpretation addresses the collapse of the wave function in a very different way. This theory suggests that the superimposed quantum state of a particle is a genuine description of that particle's physical system, but not in a probabilistic sense. Instead, the many-worlds interpretation suggests that there are *infinite parallel worlds that exist alongside our own.* It says that whenever an electron is measured as a particle and the observation shows that the wave function collapses to zero everywhere the particle is not, what actually happens is that the electron continues to exist in all other possible locations as a particle—only for those locations, it exists in parallel worlds. This means that if you have an electron wave within an atom that has the probability of being found in one of one hundred locations, and after measurement the electron reveals itself as a particle in only one of these locations, there are still ninety-nine other parallel worlds, *which literally exist*, where the electron would be found as a particle in a different location. Every possible outcome of a measurement actually occurs as the universe branches out into multiple worlds. According to this interpretation, all of these worlds are real, even though they cannot ever access each other. Everett argued that separation was no reason to reject the theory; to say that multiple other worlds do not exist just because we cannot detect them would be like saying that the Earth does not orbit the Sun because we cannot perceive the Earth's movement.

There are a number of other interpretations of quantum mechanics. One group of interpretations, of which there are a few variations, is called hidden-variable interpretations. Proponents of hidden-variable interpretations believe, as Einstein did before his death, that quantum mechanics is somehow incomplete. These theories consider the quantum state description to be a partial description of reality, and that what is missing are variables that we are currently unaware of. If we only had access to these hidden variables, then quantum mechanics would no longer be probabilistic—it would be determined. In fact, advocates of hidden-variable interpretations believe that everything is *already* determined, but that we are somehow ignorant of the variables needed to prove the determination.

How do we best talk about a level of reality that we cannot observe, and which behaves in ways we often cannot even imagine? What do we do with a picture of the universe that has nonmaterial dimensions that affect our physical reality, but which we cannot access? The wave/particle duality revealed that we cannot make such clear distinctions between matter and (alleged) empty space. Instead of being solid objects, atoms were exposed as electromagnetic waves of pure possibility. But questions remain: Where do these waves come from? What is it exactly that organizes itself into solid matter? If solid matter dissolves into a sea of mathematical potentiality, what is beyond that sea? *Where does matter actually come from?*

How can science entertain a position in which events that take place in nonempirical realms are responsible for the physical universe? The quantum behavior of subatomic phenomena has forced us to accept that the basis of our material world is nonmaterial; indeed, there are scientists who believe that these fundamental processes occur outside of spacetime itself. This means that whatever and wherever that nonmaterial region is, it does not lie at all within the matter or space that comprises our universe. However, some argue that this intangibility does not make the realm in question any less *real.* If we consider this realm to be made up of forms or processes which have the potential to emerge into our material world, then this realm holds within it *all* the potentiality of the physical universe. If there is a realm outside of our own that organizes everything within our reality, then whatever happens there is responsible for all of the events that occur within spacetime. All tangibly physical things emerge from this realm, although this realm is not tangible in itself. And if these forms give rise to what we consider real, mustn't we also consider *them* real, even though they lie outside the theater of spacetime?

Ruth Kastner, a physicist and philosopher who specializes in the transactional interpretation of quantum mechanics, argues this point. The transactional interpretation, which we will return to in more detail later, is not an observer-dependent model.

It proposes that quantum possibilities are "necessary precursors to spacetime events,"[3] even though not every possibility materializes. These "possibilities" exist outside of what we call spacetime. Kastner sometimes employs the term *quantumland* (a reference to Edwin Abbott Abbott's *Flatland*, the story of a being that lives in a two-dimensional universe who suddenly meets a three-dimensional sphere and thus learns about higher dimensions) to describe the realm in which these processes take place. According to Kastner, this level of reality is *sub*empirical, meaning that it lies outside of the boundaries of empiricism because it exists outside of our spacetime dimension, and therefore cannot be observed through our senses in the familiar scientific way. However, she argues that just because some parts of reality exist in such a way that we cannot access them, this does not make them less *real*—although she does concede that they may not be considered *actual.* According to Kastner, *actual* events are those that "exist as a component of spacetime."[4] Kastner explains that the *possible* events which take place in this subempirical realm do exist, just not within spacetime. However, it is these very processes that give rise to all the events that do occur within spacetime. These "possible" events are therefore real, but they are not "actual."

Kastner is a *realist*, which means that she considers this theory to be describing something that really exists on its own, independently of any individual's observation or consciousness. Potential events have a physical reality, even though they may not possess spacetime coordinates. In one sense, this theory is an attempt to stretch the definition of what can be considered physical. In another sense, it is a metaphysical formulation and therefore does not adhere entirely to the classical terms of science, which means that many physicists would consider the concept to be outside the bounds of legitimate consideration. But there is a long history of great physicists (Einstein, Bohr, Heisenberg and Schrödinger to name a few) who did engage in this kind of thinking, and thought doing so was necessary because the issue of where matter ultimately comes from informs *all* of physical science in a very real way.

For example, Heisenberg wrote, "Atoms and other elementary particles themselves are not real; they form a world of potentialities rather than things of the facts."[5] Here, Heisenberg acknowledged that collectively, quantum phenomena comprise a "world" in themselves. Yet, according to the Copenhagen interpretation, the external world has no objective reality other than what the observer's consciousness perceives. Heisenberg and Bohr both stopped short of attempting to visualize this pre-spacetime world as something that exists physically. Kastner, on the other hand, uses the word "real" differently from Heisenberg. To her, there is an existing part of nature, a physical part, that lies beneath our own dimension and generates the possibilities of matter, which then emerge into spacetime. Although this world cannot be seen and assessed in the usual scientific way, it *exists* and is more than just a mental concept.

In order to illustrate concepts that are radical and/or hard to imagine, humans turn to metaphor, analogy and—when a picture of reality transcends even language's ability—symbol. Although this way of thinking may not be entirely logical in a scientific sense, it becomes crucial in understanding certain necessary concepts that transgress rationality. Schrödinger defends this process in an essay written later in his life. He encourages the use of "an allegoric picture of the situation" in circumstances where "logical thinking brings us up to a certain point and then leaves us in the lurch."[6] He points out that there are hundreds of examples of this practice in science and that it has long been accepted as valid and useful.

Kastner employs a few analogies in order to illustrate the concept that there is an unobservable subtle reality underneath our own, which organizes and generates the world we see. She describes the perceptible portion of reality—meaning that which is actualized in spacetime—as the "tip of an iceberg," with the unactualized, imperceptible—but still *real*—quantum portion of reality as the part of the iceberg that is submerged under the ocean, as well as the ocean itself.

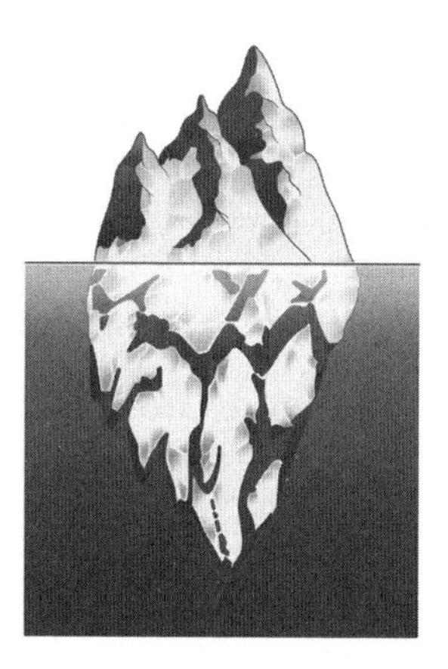

The "tip of the iceberg" observable world that we inhabit is the version of reality that is described by classical Newtonian physics. This observable world has, as its foundation, a hidden collection of possibilities that transcend four-dimensional spacetime. These possibilities are physically real, although they are not actual, and are described by quantum theory. Kastner also employs the analogy of a geode to describe this relationship in a different way. She suggests that spacetime may be compared to the crystalline structure that grows within a geode's outer "shell of amorphous material."[7] Geodes are formed when certain minerals found in water are deposited into a hollow globule of lava.

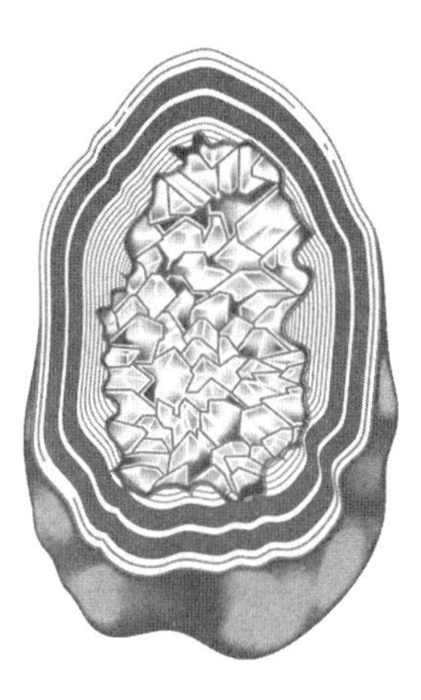

Kastner concedes that her metaphor might be more accurate if the minerals came from within the lava "shell" itself, but still her point is clear. The shell from which the crystalline structure grows in a geode is analogous to the part of the iceberg that is submerged underwater; it is the pre-spacetime dimension of unfixed possibilities. In this way, our universe is not something that is "already there." Instead, it is something that grows continuously as the possibilities inherent in the pre-spacetime outer shell become actualized. From this perspective, spacetime isn't the backdrop upon which events occur, it is instead "the structured set of events themselves."[8]

We find another advocate of a similar formula in the physicist John Wheeler. Wheeler explained that the space within the universe could not serve as the *stage* upon which the universe itself moves. He wrote, "nobody can be a stage for himself," and insisted that the field upon which spacetime "does its changing" must be much larger than spacetime itself. Wheeler proposed that this larger region be called *superspace* and that it does not possess the three or four dimensions that our reality does. Instead, this superspace is "endowed with an *infinite* number of dimensions."[9]

Schrödinger also employed the analogy of a crystal to explain a different philosophically conceivable, but irrational (at least to the Western empirical perspective) picture of reality. When remarking on the notion that reality is somehow an undivided whole and the plurality of the world may only be an illusion, which is implied in part by the nonlocality of entanglement, Schrödinger relied on the analogy of a multi-faceted crystal. If an object were to be reflected in the many facets of a crystal, it would appear that there were hundreds of images of the object, when in reality there is only one. The object itself is not multiplied, but it appears to be so to our limited human senses.

David Bohm, a protégé of Einstein's, also turned to analogy to explain entanglement, and while Bohm's theories were never accepted by mainstream science, they still provide a useful way to think about reality. Bohm theorized that reality may have two separate levels. One level would be the everyday, tangible world we live in, where Newtonian laws rule and in which two entangled particles would seem separate from each other, although they share an unbreakable connection. Underlying this level would be a deeper order of reality—a vast, potentially infinite level that births all physical objects. In the 1980s, Bohm referred to this underlying level of reality as being of an "implicate order," and referred to the reality in which we live as being of an "explicate order." The realm of implicate order was in essence a realm of pure information, from which our physical, explicate reality unfolds and then returns. The informational quality of the implicate level of reality means that it does not adhere to our three-dimensional concepts of space and time. To Bohm, all appearances of *location* disappear at that level. He proposed that this could potentially account for the phenomena seen in entangled particles—in the implicate order of things, two entangled particles are never actually separated at all. Instead, they exist together permanently in some kind of infinite informatic unity, and this in turn implies a certain wholeness to reality.

To support his theory, Bohm turned to the idea of a hologram. Holography, which was at the height of its popularity in the 1980s when Bohm was working on this theory, uses the wave nature of light to create three dimensional images of objects. Unlike traditional photography, which might use a lens to focus on an image and then imprint that image onto a piece of film by registering where light is and isn't, holography records the shape of a light wave after it bounces off a particular object. By using the information contained in the interference patterns caused by intersecting light waves as they bounce off an object, holography is able to create a three-dimensional image. The theory behind this technology was accidentally discovered by a scientist named Dennis Gabor in 1948. While he was working on improving the image resolution of electron microscopes, Gabor stumbled upon an idea that he first called "wavefront reconstruction" (the wave in question being light waves). He coined the term "hologram" to describe the product of such reconstruction, the Greek roots of which break down into *holos*, meaning whole and *gramma*, meaning message.

Modern holography uses laser light, which is a consistent and pure form of light and is therefore particularly good at creating strong interference patterns. A single laser beam is split into two beams by a device called a beam splitter. The first beam, the "object beam," is then bounced off of a mirror to illuminate whatever object is being recorded (say an apple). The reflected light is then recorded onto a photographic plate. The second beam, "the reference beam," is directed at the photographic plate and collides with the object beam. The interaction between these two beams creates interference patterns, much like those in both of the double-slit experiments.

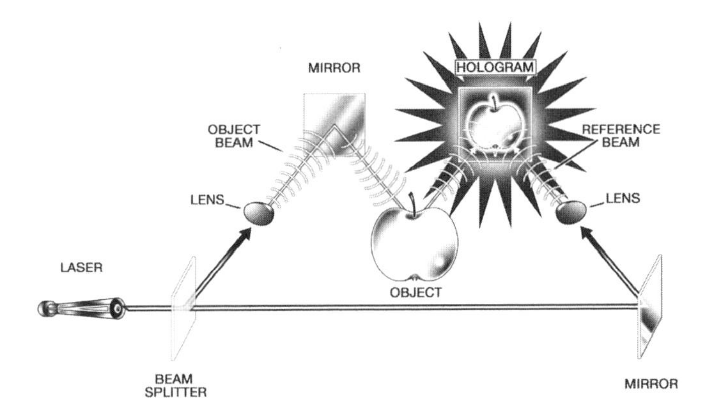

What gets recorded onto the holographic film or plate then is an interference pattern that looks like speckles. If you were to look at a developed holographic plate with the naked eye, you would only see this pattern of light wave interaction. However, when you shine a laser through it, a three-dimensional image of the object appears.

One of the most interesting things about holograms is what happens when you cut a holographic film in half. When you cut holographic film in half, and then shine a laser through each half, instead of seeing half an image of whatever object was recorded, you will instead see *two whole images* of the object in question. So, if you halved a piece of holographic film with the image of an apple encoded in it, and you shined an appropriate laser through both pieces, you would see two whole apples instead of two halves of one apple.

This is because every point on the entire film exposure of a hologram contains *all of the information* about the image being produced. The interference patterns encoded in a specific piece of film contain *the whole in each part*. Each time you reduce the film in size, you will still get the whole image of whatever was first encoded, although it will become fuzzier and fuzzier the farther you go.

Bohm eventually found the concept of a hologram itself to be too confining—namely because a hologram requires a fixed level of deeper reality. Bohm was convinced that reality as a whole was not static and therefore the implicate order itself could not be stable or fixed. However, he did retain the idea of the relationship between the two levels of reality as being holographic, and employed the hologram as a useful analogy. He called the relationship that joined the two levels "holomovement" or "holoflux." Bohm saw correlations between the wave/particle duality of matter and the way that information is stored in a hologram. Both are dependent on the behavior of interference patterns and both suggest that an object that appears solid might not actually be so. In addition, because an electron passes through the two slits of a double-slit experiment as a wave but arrives at the back screen as a particle, Bohm suggested that the electron's nature was, in some large way, dependent on the context of its environment. To Bohm, this implied that the electron does not actually have "its own separate nature entirely, but is (instead) internally related to the whole." He also saw entanglement as an issue of "quantum nonlocality," which could be resolved by this kind of holographic thinking. Because, under particular conditions, you can find particles that are intimately connected through entanglement even though they are physically separated, this suggested to Bohm that certain sets of particles are somehow united into a whole "that is not reducible to [the] action of its parts."[10]

Like Schrödinger's crystal, Bohm's holographic picture of the world implies that reality may be some kind of multiplicitous projection of a deeper reality, which is itself an unbroken whole. The shared importance of interference patterns in the wave/particle nature of reality and holograms seems to bolster Bohm's analogy, although the hypothesis has since been all but discarded by mainstream science. Nowadays, holographic analogies are utilized mostly by string theorists to describe certain properties of quantum gravity. However, in terms of trying to envision a dimension beyond spacetime from which our reality emanates, Bohm's analogy is invaluable, and there are certainly other aspects of nature that suggest a holographic structure—like the human genome. The human body is composed of trillions of cells, but not all cells are the same. There are immune cells and skin cells, and cells which comprise neurons or our bones. However, almost all of these cells contain a nucleus with the same DNA base pairs. Although these cells grow into different parts that carry out different functions, each contains the same coded information from which, in theory, not just a new cell but an entire new being could be created.

Another example of reality's potential holographic structure can be seen in the nature of black holes. Because a black hole possesses a tremendous amount of gravity, it used to be thought that not even light could escape its pull, and that whatever got sucked into one would vanish. However, recent mathematical theories suggest that this might not be so. It turns out that if you were to throw your phone inside of a black hole, even though the three-dimensional object of the phone would disappear, a two-dimensional copy of the information that constitutes the phone's physical nature would get spread out over the surface of the black hole. This means that the informational content of whatever disappears into a black hole gets expressed, in its entirety, on the black hole's exterior. In this way, the phone can be said to exist in two places—the three-dimensional object that is inside of the black hole and a two-dimensional informatic version that survives on the surface. This suggests that you could, in theory, reconstruct your lost phone using the information left behind. Some physicists have hypothesized that our universe may behave in a similar way. The three-dimensional world that we know might possibly be a projection of information stored on some two-dimensional shell that surrounds the entire universe (not unlike the geode analogy advanced by Ruth Kastner).

Interestingly enough, there is a Buddhist system of philosophy, which appeared in China in the seventh century, which relies on a mythological image very similar to Bohm's holographic universe. In the Huayan school of Buddhism, all of existence is regarded as being an infinite web of interconnection. If one strand of the web is excited or disturbed, the entire web will be affected. The Huayan school, which although Chinese in origin is based on Indian Buddhist views, takes as its central text the *Avatamsaka Sutra*. In classical Sanskrit, *avatamsaka* means garland or wreath and the sutra (Buddhist scripture) characterizes our reality as a universe of infinite realms upon infinite realms, each of which contains all others. A mythological image that is often referenced in the Huayan school to illustrate such a fractalized cosmos is that of Indra's Great Net of Gems. The story behind the image is that within the divine palace where Indra, the king of the Devas, resides, there hangs a wondrous net that extends unendingly in all directions. On each node of the net, where its brilliant strands cross each other, hangs a single sparkling gem. Because the net itself is infinite, there are infinite gems. The most spectacular thing about Indra's net is that each singular gem contains the reflection of *all the other gems in the net.* Furthermore, each of the infinite gems that are reflected in a single gem in turn reflect all of the other gems in the infinite web, revealing an inexhaustible process of reflection. Therefore, to look upon one is to look upon the whole.

The Huayan universe is nonteleological, meaning that there is no concept of a creator or definite purpose to human life. There is also no theory of time. Reality exists of itself, and adheres to a physics all its own. Things that seem separate are in fact, interrelated, and what affects one part of the universe will, by nature of its interconnectedness, affect all other parts. What is considered true is not physical objects or individual minds, but the *relationship* between all events and organisms. Like Bohm's understanding that the electron is dependent upon its environment, and therefore is not separate from it, the world described by the Huayan school derives its existence from its inseparable wholeness. Like Bohm's universe, the Huayan universe is ever-changing and non-static. The ability for one jewel in Indra's net to reflect all others, even those that lie on the other side of an infinite cosmos, suggests a kind of nonlocality that also implies that our three-dimensional world is an illusion, one in which we mistakenly take the reflections of reality to be reality itself.

Much has been written about the supposed relationship between various Eastern religions and philosophies (namely Hinduism, Buddhism and Taoism) and quantum physics. The best-known books on the subject are perhaps *The Tao of Physics* by Fritjof Capra and *Mysticism and the New Physics* by Michael Talbot. Both books have significant value in exploring the correlates between Eastern philosophies and the implications of quantum mechanics. Various Eastern religions and philosophies had an important influence on the founders of modern physics, and the convergences between the two different worldviews provides many opportunities to make meaning.

The ideas about reality that accompanied the discovery of subatomic behavior were unthinkable to the Western scientific mind. However, various Eastern religions and philosophical schools had been thinking about the unthinkable for centuries at that point, with much emphasis placed on the literal existence of regions of reality that lie outside of space and time. It is well documented that the quantum physicists of the early twentieth century knew of and turned to these philosophies for inspiration and for solace, but it should be noted that most of them did not attempt to draw exact correlations, nor did they think it prudent to do so.

Still, direct correspondences can indeed be found between Eastern philosophies and the worldview ushered in by quantum mechanics. A perspective shared throughout both is that the three-dimensional world is not the full picture of reality, and that the full picture of reality is nearly impossible to imagine. In a subset of the Vedanta school of Hindu philosophy called Achintya-Bheda-Abheda, reality as a whole is regarded as unthinkable. *Achintya* translates as inconceivable, *bheda* translates as difference, and *abheda* translates as non-difference. Reality is reality beyond comprehension, beyond expression. In another Hindu formulation, the concept of a shapeless, omnipresent, invisible, transcendental, immanent and ultimate reality in which there is no time or space is called Brahman. The world of matter is considered an illusion called māyā, which entangles consciousness and does not permit humans to witness the actual truth of existence. The concepts of Nada and Bindu from the Upanishads, which are a series of seminal religious-philosophical Hindu texts, have been compared to the wave/particle duality. Nada is sometimes explained as cosmic vibration, and is described as the sound heard during deep meditation. Bindu means a dot or point, and references a specific coordinate at which the energy of Nada is collected into a physical point. Recently, in 2023, researchers from the Sapienza University of Rome and the University of Ottawa in Canada used a holographic imaging technique (called off-axis holography) to image quantum-entangled photons. The images are striking: the entangled photons make a perfect yin-yang.

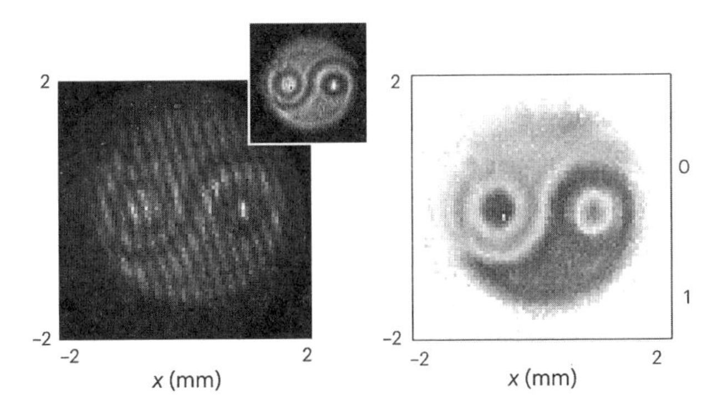

(Image credit: *Nature Photonics*, Zia et al.)

These are just a few of the many equivalences that have been noted. However, the influence that various Eastern philosophies has had on quantum physics has been more nuanced and dynamic than just drawing correlations. David Bohm, the scientist mentioned earlier whose work included developing theories of nonlocality and the wholeness of reality, considered Eastern religion an influence on his life since childhood and was friends with the Indian philosopher Jiddu Krishnamurti. J. Robert Oppenheimer, the father of the atomic bomb, was greatly interested in Hinduism and learned Sanskrit while teaching at Berkeley in the 1930s. Oppenheimer famously quoted a line from the Bhagavad Gita, "Now I am become Death, the destroyer of worlds," after the first successful nuclear explosion. The line refers to the pure destruction of reality and time, no matter what humans do or do not do. Schrödinger was very vocal about the impact that Vedantic philosophy had on his views of reality and consciousness, especially towards the end of his life. He was also guided significantly by the work of Arthur Schopenhauer, arguably the first major Western thinker to be influenced by the Upanishads. Schrödinger regarded the multiplicity of matter and consciousness to be an illusion and considered the "doctrine of the Upanishads" to be correct: All plurality was only a matter of appearance, all reality was, in fact, one. However, Schrödinger was also extremely cautious in recommending the admission of any kind of religious views into science, because he did "not want to lose the logical precision that our scientific thought [had] reached," which he considered "unparalleled anywhere at any epoch."[11]

In 1929, Heisenberg met with the Indian poet and philosopher Rabindranath Tagore, who introduced him to Indian philosophy. From that meeting, Heisenberg learned that the impermanence of matter, the concept of relativity and the interdependence of all things—concepts that had baffled him and other physicists—were the very foundation of Indian religious and philosophical traditions. He said that after speaking with Tagore, "some of the ideas that had seemed so crazy suddenly made more sense. That was a great help for me."[12] Five years before he died, Heisenberg wrote the following in a letter to Fritjof Capra: "The kinship between the ancient Eastern teachings and the philosophical consequences of the modern quantum theory have fascinated me again and again."[13]

Bohr had an interesting relationship with the yin-yang, the Taoist symbol that symbolizes the dynamism of nature, the Tao.

The Tao contains two equal and complementary parts: Yang, which is considered active, positive, masculine and related to the Sun, and yin, which is considered passive, negative, feminine and related to the Moon. This relationship is present within every object and event in the universe, from living beings to inanimate matter and even forms of energy. The magnetic tension between yin and yang is called qi. Qi can be visualized like this: if you were to take a little piece of metal and place it between the positive and negative poles of a magnet, it will vibrate for a moment and then come to rest at the position where the magnetic tension is most acute. That point is qi. The Chinese worldview that evolved from this concept of Tao is that the universe exists in perpetual flux. Nothing stays still, nothing is static. Although there are tendencies, relationships and cycles of nature that can be witnessed, everything changes all the time, at every moment.

Historians of science debate the extent to which Niels Bohr's scientific work was directly influenced by the yin-yang. However, the similarities between wave/particle duality and a symbol that depicts the complementarity of nature as two mutually dependent opposites are plain to see. A feature shared by both concepts is the presence of one aspect in the form of the other. In the yin-yang there is a small light dot embedded in the dark section and a small dark dot embedded in the light section. This means that the seed of the dark energy lies within the light energy and the seed of the light energy lies within the dark. Likewise, every particle contains the potential of a wave and every wave the potential of a particle.

It remains unknown if or how much Bohr was influenced by Taoist philosophy during the construction of his quantum theories, but he did become very linked with the yin-yang later in his life and after his death. In 1947, the King of Denmark Frederik IX conferred upon Bohr the Order of the Elephant, Denmark's highest-ranked order of chivalry. Bohr designed his own coat of arms for the honor and included the yin-yang symbol as well as the Latin phrase: *contraria sunt complementa*, which means "opposites are complementary." A bust of Bohr on the campus of the University of Copenhagen also features a yin-yang.

We do know that Bohr considered complementarity to be the only objective description of nature. To Bohr, nature could only be described in a relational manner. Reality is not represented in one state of matter or the other. Instead, reality itself lay in matter's dual nature and its ability to be expressed as either a particle or wave. We also know that Bohr was aware of the correspondences between quantum theory and various Eastern philosophies, and even suggested it necessary to reference such philosophies when trying to understand certain aspects of quantum thinking. In a speech given in Bologna, Italy, in 1937, Bohr told his audience that a "parallel" to "the lesson of atomic theory" can be found in the kind of "epistemological problems with which already thinkers like the Buddha and Lao Tzu (an ancient Taoist philosopher) have been confronted when trying to harmonize our positions as spectators and actors in the great drama of existence."[14]

Ruth Kastner also turns to the yin-yang symbol to describe something inherent in quantum theory. Kastner, however, does not reference the symbol to describe the complementary nature of the wave/particle duality. Instead, she uses it to explain a central facet of the transactional interpretation of quantum mechanics. As noted above, Kastner believes that quantum possibilities exist in a physical dimension outside of spacetime. But just what are the processes that take place in such a realm? According to the transactional interpretation, spacetime events are created when a quantum system (like an electron wave) "travels from a source, called an *emitter*, to a destination, called an *absorber*."[15] Unlike other quantum interpretations, the transactional interpretation does not consider quantum interactions to be sufficiently described in terms of an entity just "being created or emitted." Instead, there must be *something* that receives whatever momentum or charge is transmitted. In the transactional interpretation, all quantum objects (like electrons and photons) are "fundamentally wavelike." The subatomic objects do their preliminary "exploring" as waves and "it's only in the very final stage that a particle-like behavior emerges."[16] A quantum object begins as an "offer wave," but without anything present to receive the offer wave, no particle will come to be. Accordingly, there is an additional process that needs to occur before the offer wave can actualize into spacetime as a particle—*absorption*. An absorption occurs when an offer wave meets what is called a *confirmation wave*, which is something like a negative mirror image of the offer wave. The energy coming from an offer wave is considered positive and is directed toward the future. The energy coming from a confirmation wave is considered negative and is directed toward the past.

This theory might sound crazy, but remember, the transactional interpretation considers these events as occurring in regions of reality that do not have to behave according to our linear experience of time. The transfer of energy from one quantum system to another can only occur if there is both something emitting the energy and *something actively receiving it.* The conventional quantum interpretation maintains a unilateral view of radiation, and says that an emitter is the sole contributor of energy and that an absorber just passively receives whatever energy has been sent its way. However, in the transactional interpretation, the receiver must be active as well—the receiver also produces a wave, although one considered as being composed of negative energy.

Kastner explains that the reason many physicists dismiss the idea of a confirmation wave is because the confirmation wave would be made up of negative energy, which means that it is "unphysical." It cannot exist within spacetime. However, we have already established that Kastner champions the existence of quantumland, a region outside of spacetime where these transactions take place. Furthermore, Kastner suggests that the ordinary quantum equations already contain aspects of the wave system that are unphysical. One example is the amplitude of a quantum wave. The amplitude of a wave is the measurement of how far the wave rises above and how far it falls below its origin point. The amplitudes of the waves underlying quantum systems are described by complex numbers. Complex numbers are number systems that are made up of both real and imaginary numbers. Imaginary numbers do not have a tangible value, and because they are used in part to describe the amplitude of quantum waves, there is something about the waves themselves that cannot be thought of as tangible. Kastner argues that this means that a quantum wave system cannot be considered as existing fully within spacetime.

For Kastner, the process of absorption is perfectly represented by the symbolism inherent in the yin-yang. The offer wave from the emitter accords with the yang principle because it initiates the action and gives off positive energy. The confirmation wave from the absorber accords with the yin principle because it accepts the energy by actively receiving it. Kastner accuses Western science of emphasizing "yang to the exclusion of yin," and says that this fails to account for a complete interaction. She compares the situation to growing daffodils where the bulb must be "*received* in the ground."[17]

The interaction between the earth and the daffodil bulb is necessary to the growth of the flower. The bulb itself, like the emissions from a quantum wave, are only half of the equation. Taken as a whole, the yin-yang-like process of emitting and receiving provides a sufficient explanation for the behavior of subatomic particles. When the subatomic entity is in its waveform, it is still a *possibility*, and this possibility is described by the interactions between offer waves and confirmation waves. When this interaction between the two kinds of waves becomes a successful "transaction," the subatomic entity appears as a particle and becomes a component of spacetime. According to the transactional interpretation, these transactions take place within a reality that is independent of any human consciousness, and therefore an observer is not needed to bring reality into existence. Matter comes to be as a result of relationships that take place outside of spacetime, in immaterial realms we cannot access. Our world, and everything in it, is the result of processes that take place somewhere beyond observable reality.

There are scientists who would take the deconstruction of matter even further. We have engaged with some physicists who conclude that matter is created from some immaterial process. However, there are others who assert that matter *does not exist*, only the process does. In a speech he gave toward the end of his life, Max Planck said "there is no matter in itself." He prefaced this statement by reminding his audience that he was a sober scientist, who had devoted his entire life to physics, and that his conclusion was drawn directly from his research on the atom. He continued: "All matter arises and exists only through a force that vibrates the atomic particles and holds them together to form the tiniest solar system in the universe."[18] Hans-Peter Dürr, a physicist who was considered a leading follower of Heisenberg, came to a similar conclusion. Dürr studied matter for over sixty years and remarked that many people would often ask him what the most interesting thing about his work was. He would tell them that the most interesting thing was that matter did not exist. Dürr said that he saw that this was evident in some ways from the beginning of his career, but he did not believe it then—however, he later understood this was the only reasonable conclusion he could draw.

Dürr said that classical physics asks "what exists" and then "how does it behave." He cautioned, however, that modern physics cannot ask that question. According to Dürr you cannot ask what is, you can only ask *what is happening?* In fact, he said that he did not call atoms atoms, but instead calls them "haps," which is short for "little happenings." This is because what we call a thing is not really a thing, *it is a process*. And a process by definition is nonstationary; it has parts and movement, and the whole is only defined by the parts and movement which make it up. Something happens, and then something else happens, and then something else happens. According to Dürr, this is a much more accurate description of matter than the concept of a fixed thing. And it is the *in between* within the process that constitutes the realness of reality. Dürr proposed that the best way to discuss reality is through intransitive verbs like living, loving, moving, being. All of reality is dynamic, not just the process by which matter comes to be. Matter, even if it looks inert and dead, is dynamic and creative; "it is alive," said Dürr. This is because all matter in our reality is *constantly* and *continuously* being generated from an unknown source. Dürr likened this source to software and the reality we know to a printout dictated by such software. Time is not something that carries dormant material along a linear current to an end. Instead, time means that, at each moment, our reality is created anew with the memory of what came before it. Dürr summed up this concept beautifully when he said "There [was] no big bang... it bangs all the time!"[19]

Dürr considered space to be completely full of matter or the potential for matter—a plenum, not a vacuum. Modern quantum field theory, which evolved out of quantum mechanics, suggests something similar. Quantum field theory says that everything in the cosmos is actually made up of quantum fields; plastic, liquid-like entities that exist throughout the universe and sometimes behave as waves and sometimes as particles. According to quantum field theory, electrons themselves are not elementary objects, but are instead little undulations of something called the electron field, which is spread throughout the universe. There are different fields for different subatomic entities and all particles are in fact just little ripples in these fields.

While Dürr might have agreed with the concept that "empty space" itself is full and generative, he might have disagreed with the instinct to separate any entity from the whole. Dürr, who also referenced the Advaita Vedanta school of Hinduism, believed that all of reality is made up of one single substance, and our separateness from that substance is an illusion. Whereas Democritus argued that the atom is "that which cannot be cut," Dürr might have said it is reality itself that cannot be cut. However, he agreed that to act in the world one must break it up into pieces. Dürr defined the observer as a construction of consciousness that does just that: treats the world substance as something that can be divided and separated. Our universe is created as a whole in every second from some unobservable matrix, all the time, in real time, so that time stretches out into itself until it does not exist. According to Dürr, this process becomes smeared out at our human level of reality and because of this, the laws of classical physics apply to us, even though our reality at its base possesses an unbreakable wholeness. What we call matter is the result of trying to make sense of our blurred-out position in the universe. Dürr refers to the kind of Cartesian dualism that engenders such an illusion as "an infection."

The physicist Carlo Rovelli also proposes that we should not think of the matter in our world as a *thing.* Instead, he suggests we think of the world as being made up of *events.* The events of the world are not permanent, but undergo unending change and transformation. Even something as "thinglike" as a hard rock is actually "a complex vibration of quantum fields, a momentary interaction of forces." According to Rovelli, a rock is nothing but a very long event, it is a process that is somehow able to maintain its shape briefly, "before disintegrating again into dust." If we were to think of the world as comprised of things, Rovelli wonders what we would consider these things to be. We know that atoms are made up of mostly empty space, and yet can be broken down into smaller particles. According to quantum field theory, these particles are agitations of particular fields. And these fields in turn are no more than "codes of a language with which to speak of interactions and events."[20] Rovelli points out that physicists have long sought the existence of a primary substance but the more that reality is studied, the less it can be described in terms of what *is*. Instead, he declares that the nature of the world seems to be much more comprehensible in "terms of relations between events."

Reality itself may be beyond comprehension. It is an ever-blooming flower. Even *spacetime*, the determined boundary of our physical existence, is not itself a physical thing. Spacetime is a consolidated dimensional description of the quantitatively determinable aspects of space and time. The concept was introduced in 1908 by Hermann Minkowski in order to reformulate Einstein's theory of relativity. Einstein himself cautioned against thinking of spacetime as something physical, saying "space and time are modes in which we think, and not conditions in which we exist."[21] It is important that we do not confuse our representations of reality with reality itself. Representations are crucial because the processes of nature reveal that reality is, in some significant way, beyond the grasp of direct human intelligence. But we should not mistake the representations for a fixed and faithful reflection of reality itself.

And yet we see the immense power of these representations. Matter does not exist; it is a myth constructed in the face of the inconceivable truth of reality. Without knowing it, the atomists of ancient Greece created a mythology that would masquerade as truth until the twentieth century. Even today the solidity of the material world seems irrefutable to our senses. Yet, as we have discussed here, quantum science has exploded the appearance that reality is a closed universe made of solid matter and empty space. We are left with a bewildering range of choices as to what is and is not real. We can say that nothing is real; not the dimension we live in, nor the one from which our material world emerges. Or we can say that they are both real. Or we can say that the realm that exists outside spacetime is real, but our physical dimension, with its solid matter and empty space, is not real. But it does not seem like we can say that our material dimension is real, but the other realm—the one of Dürr's software, Kastner's wave transactions, and Bohm's implicate order, the one that comprises the part of the iceberg submerged underwater, or that makes up the geode's shell—is not real. The formulation that something emerges from nothing calls into question both the reality of *something* and the boundaries of *nothingness* itself.

In part one of this essay, we have explored a tiny subsection of the ways in which matter, and its negation, have been expressed in physical sciences and related metaphors. But can we know the true essence of matter outside of the consciousness that ponders it?

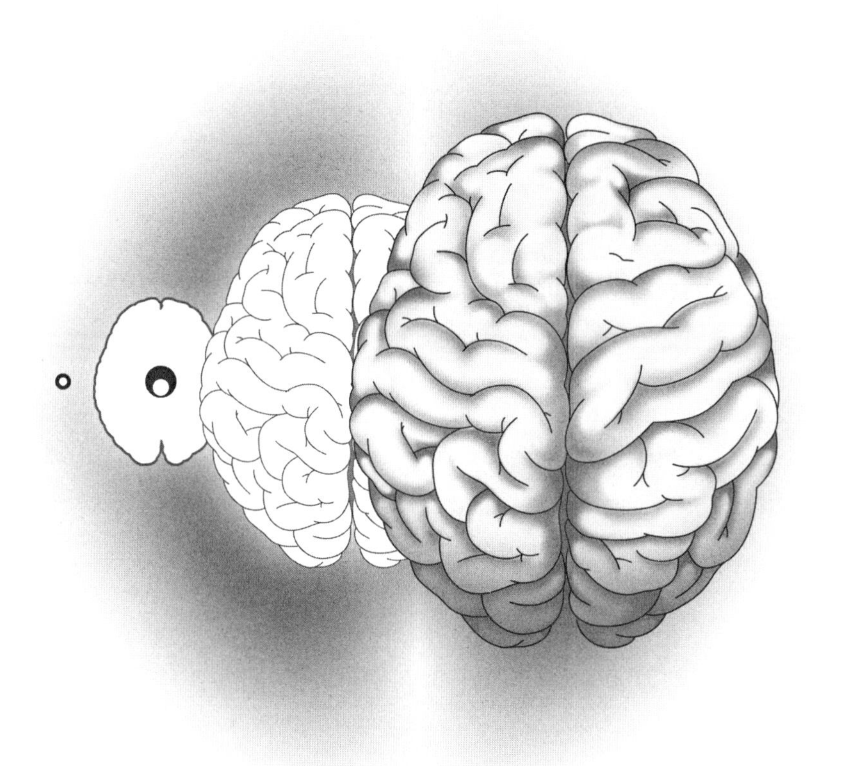

The Myth of Matter

PART II

The mind is the matrix of all matter.

—Max Planck

Since psyche and matter are contained in one and the same world, and moreover are in continuous contact with one another and ultimately rest on irrepresentable, transcendental factors, it is not only possible but fairly probable, even, that psyche and matter are two different aspects of one and the same thing.

—Carl Jung

What is the relationship between the physical world and the world in our head? Does matter actually ever stop somewhere and consciousness begin? Throughout history, there have been countless attempts to answer these questions from scientific, philosophical and religious points of views. Theories abound from each school of thought about the specific nature of mind and matter and the ways they are connected. The dualism that engenders the split between them, so that mind and matter are separate entities to discuss both as things in themselves and as things in relation to each other, is itself a theory—although Western science tends to treat it as doctrine. Made famous by French philosopher and scientist René Descartes in the seventeenth century, *mind-body dualism* is the view that mind—and by extension consciousness—is nonphysical, nonspatial and categorically different than body. This dualistic view gave rise to what is known as *the mind-body problem*, which seeks to answer the question of how and in what ways are physical and mental phenomena related. As it stands in its modern form, the questions usually fall into two categories: Are mental states fully separate from physical states? Or is the mind an epiphenomenon, that is a secondary product or phenomenon, of material processes that take place in the brain?

However, the fact that these are the two big questions that get asked does not make them the *right* questions to be asking. *Materialism* is the philosophical theory that matter is the foundational bedrock of reality. However, privileging matter as reality's primary substance, and asking how consciousness relates to matter (rather than taking consciousness as the starting point) is a historical circumstance that can be both traced and questioned. While modern mainstream science is materialistic and upholds the position that consciousness is a byproduct of the physical processes of the brain, there is a robust counterflow of philosophical reasoning that argues otherwise.

The analytical psychologist Carl Jung had a unique formulation regarding the relationship between mind and matter. Although Jung's primary concern was *psyche*, "the totality of all psychic processes, conscious as well as unconscious,"[1] because psyche and matter exist in the same reality, and are both present in each and every human being, Jung was also concerned with the nature of matter.

And for Jung, the nature of matter was deeply connected to the nature of the *collective unconscious.*

Jung first wrote about the unconscious in a 1912 book called *Psychology of the Unconscious*, and first wrote about the collective unconscious in a 1916 essay titled, "The Structure of the Unconscious." The psychoanalyst Marie-Louise von Franz, who worked closely with Jung, points out that both Jung and his mentor (and later rival) Sigmund Freud "discovered"[i] the unconscious at the very same time as the advent of quantum mechanics. While Freud mainly viewed the unconscious as a boundaried, personal repository of repressed drives, Jung saw it as significantly more than that. In addition to the kind of personal unconscious that Freud described, Jung argued for the existence of the collective unconscious—also called the objective psyche—which transcends the boundaries of individual consciousnesses, is not conditioned by personal experience, is shared by all of humanity, and contains the *archetypes* as its organizing principles. According to von Franz, Freud's focus on the "drive aspect of the unconscious" was intended to align his work with the mainstream medical knowledge of that era, such as "brain physiology, endocrinology, and research on general biological processes altogether." Jung, on the other hand, did not feel the need to argue for "premature equivalents between the unconscious and physical and material processes." This was not because he didn't believe in such equivalents—he did—but he believed that psychological phenomena should first be investigated within its own realm, the psychic realm, before attempting to create direct links between psychic and somatic processes. By taking this position, Jung was already countering the "material prejudice of his time," which, like most mainstream science today, favored the idea that consciousness (and by extension psyche) is an "epiphenomenon of physiological processes."[2] Jung held the position that as research into both somatic and psychological phenomena continued, the link between them would reveal itself naturally.

According to Jung, the aspects of psyche are divided into three main subgroups. There is the *ego* or *personal psyche*, which encompasses absolutely everything known or felt by an individual person at a specific moment. There is the *personal* or *subjective unconscious*,

i Von Franz clarifies that while there had long been discussions about the existence of the unconscious, Jung and Freud were the first to prove its reality through empirical investigation and experiments.

which, along with Freudian repression, contains everything a subject knows but is not thinking at a specific moment, everything that was once known but has since been forgotten, everything perceived by the senses but not recognized by the conscious mind, everything that is felt, thought, remembered, desired or done involuntarily or without direct attention being paid to it, and all the thoughts, feelings, desires, or actions that lie in the future and will eventually "come to consciousness."[3] The personal unconscious also contains *complexes*, which are emotionally charged "content clusters that form associations around a nuclear element" and draw "ever more associative material to themselves."[4] And then there is the *collective unconscious* or *objective psyche*, which is a stratosphere of psyche that is shared by all of humanity, and is "made up essentially of *archetypes*."[5]

Archetypes are primordial patterns that act as "templates for psychic activity."[6] Archetypes are both objective universal psychic structures that influence the emergence of specific images into consciousness, as well as dynamic agents that constellate spontaneously according to their own laws. They are often equated with physiological *instincts*. Although archetypes themselves are unobservable because they lie within the collective *un*conscious, we come to know something about them through the thoughts, fantasies, ideas and emotions that they produce. Therefore, there is a necessary distinction to be made between the archetypes themselves, which are irrepresentable, and the archetypal *image*, which emerges due to "some inner or outer state of need."[7] By virtue of the fact that archetypes make up much of the collective unconscious, and organize themselves into various content that eventually emerges into consciousnesses, archetypes are considered *a priori*, meaning that their existence is self-evident and not based on anything that came before them. Von Franz suggests that they are very much like the *a priori* elements of material reality, such as the speed of light or the duration of half-lives.[8]

Although we cannot know archetypes themselves, archetypal images are undoubtedly familiar. The *child* is the archetypal image of potentiality, innocence, naivety and future maturation. The *persona* is the archetypal image of the public-facing "mask" one wears to fit in with society and interact with social groups at large. The *hero* is the archetypal image of the quest-oriented journey that seeks to transform either the hero themself or society at large through the successful completion of challenging tasks. In addition, events like *birth, death* and *marriage* are also considered archetypes.

Jung turned to the presence of common archetypal images found throughout world religions and mythologies to argue for the existence of the collective unconscious. He argued that, because archetypes are shared by all mankind, the fact that there are structurally similar archetypal images in myths throughout world cultures points to the existence of a primordial realm of psyche that preconditions much of our psychic existence. (It is important to make a brief note about language here. As skillfully pointed out by philosopher and computer scientist Bernardo Kastrup in his book, *Decoding Jung's Metaphysics: The Archetypal Semantics of an Experiential Universe*, Jung used the term "psyche" in various ways throughout his writing to mean different aspects of psyche as defined above. Therefore, as we continue, when there may be confusion about which aspect of psyche [i.e., whether psyche denotes the entirety of the psychic realm, or conscious aspects of personal psyche, or the objective psyche/collective unconscious], clarity will be provided in adjoining parentheses. Similar clarification will be provided for the term "spirit," which Jung also used to mean different things in different contexts.)

In order to think about psyche as a whole, Jung employed the analogy of the spectrum of light, from infrared through visible light and beyond to ultraviolet.

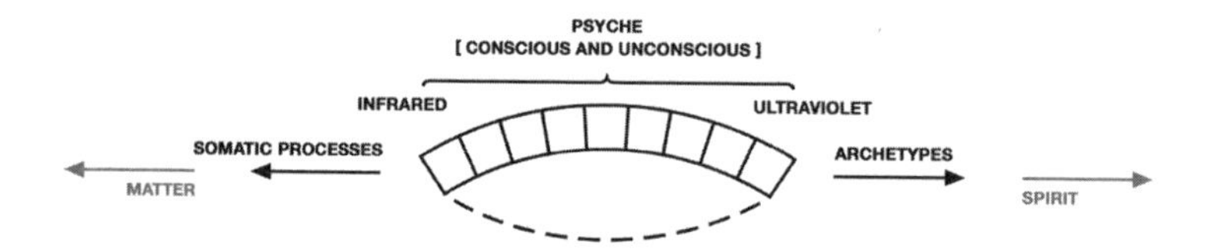

Archetypal images and thoughts lie on the right side of the spectrum, at the ultraviolet pole, while the infrared pole on the left represents the point where "psychic processes flow or merge into instinct and physical processes."[9] The middle zone of the spectrum, corresponding to light we can perceive, is the realm of personal psyche (ego consciousness). Von Franz argued that only in the middle zone is there a certain level of freedom possible for an individual, because if an individual ego gets too close to either pole, that ego is then subjugated to the control of unconscious complexes or drives. For instance, Joan of Arc might be considered to reside somewhere closer toward the ultraviolet end of the spectrum, as she was possessed by spiritual or religious motivations or ideas, while someone struggling with food or sex addiction might reside closer to the infrared end of the spectrum, being possessed by drives or compulsions linked to the body. Only in the middle region can the ego participate voluntarily in psychic processes, while at either end there is a kind of automatism. The ability of either pole to influence the personal aspects of psyche is mutual, as seen when certain chemical changes in the body alter psychic states (as in drugs) or when certain psychic states cause chemical changes in the body (as when anxiety induces a rash).

It is through this analogy that we begin to see what Jung truly thought of matter, as well as its relationship to psyche. The imperceptible archetypal structures that emerge into psyche as archetypal images lie at the ultraviolet end of the spectrum, and it is here that sudden inspiration, epiphany or spiritual insight spontaneously appear. Von Franz also argues that it is here where parapsychological phenomena occur. From the ultraviolet end of the spectrum, psyche eventually passes over into *spirit*, which von Franz defines as the "dynamic aspect of the objective psyche (collective unconscious)." It is spirit that inspires and moves our thoughts and emotions, that "spontaneously produces and orders images in the inner field of vision," and that is also "the composer of dreams."[10]

At the infrared pole, psyche merges with the body of an organism and participates in its functions and symptoms. It is at this end of the psychic spectrum where psyche merges with physiology and participates in a body's "chemical and physical conditions." From there, psyche passes into what Jung called the *psychoid*[ii] which is the part of the psychic system that can *never* be known to psyche proper (our ego consciousness or personal unconscious), and is most often used as a term to describe where the "psychic element appears to mix with inorganic matter."[11] What this means is that, from the infrared end of the spectrum, *psyche eventually passes over into matter*.

Although never explicitly stated by Jung, it may be useful to consider matter lying at the point of the spectrum where infrared light becomes microwaves, and spirit lying at the point where ultraviolet light becomes x-rays. According to Jung, reality is a unitary substance where matter and spirit stand as two poles in opposition to each other, and we as humans can only know of either, as psychic experiences, from our position in the middle. As Kastrup put it, "For Jung, there is no spirit separate from matter and no matter separate from psyche."[12] As seen by the very nature of the light spectrum analogy, reality is ultimately only one substance; just like there are colors on the light spectrum we cannot see but are still fundamentally optical radiation, so too are there aspects of the reality substance that are unknowable to us, but are nonetheless a part of the substance.

To Jung, psyche proper (our personal consciousness and subjective unconscious) is privileged as the theater of existence. We can never know what spirit, as defined above, is in its "trans-psychic essence,"[13] and although we interact with matter via our senses and sense-enhancing instruments, we can also never know what matter is in its trans-psychic essence. Both spirit and matter are not accessible or understandable in their inherent nature, but are "ultimately forms of appearance of essentially transcendent (that is trans-psychic) being." The only reality we ever know as humans is the psychic reality composed of the immediate contents of our consciousness, which we then either ascribe to "a material or mental-spiritual origin."[14] This means that even though matter is a physical category of the sciences, it is also a psychic category that can be investigated by philosophy and religion. Both matter and spirit are, in essence,

---

ii Formally, Jung used the term psychoid to describe aspects of psyche (as a whole) that *never* cross the threshold of consciousness and are truly and completely unknown, so as psyche passes over into spirit, it would also pass through a psychoid region. However, he used the term most often to speak specifically about "the part of the psychic realm where the psychic element appears to mix with inorganic matter." (von Franz, *Psyche & Matter*, 4)

working hypotheses that humans have invented to express certain contents of consciousness.

If matter and spirit are to be seen as two ends of a spectrum, and therefore two expressions of the same (ultimately psychic) substance, then the idea of matter as a separate entity with a fundamental nature different than psyche is indefensible. In this formulation, matter does not exist in any way we can claim true knowledge of, at least not in our inherited Western materialist tradition. Matter is not separate from our experience of psyche, and there is no definable boundary between the contents of our consciousness and matter as something outside of our consciousness.

In another book by Bernardo Kastrup, *Science Ideated: The Fall of Matter and the Contours of the Next Mainstream Scientific Worldview*, the philosopher and computer scientist presents a refutation of materialism that works well with Jung's ideas. Kastrup argues that the entire enterprise of modern science, as it dawned in the seventeenth century, was built on "perceptual experience" defined by "sensed qualities." Early scientists ascertained the reality of the phenomena they investigated through sight, touch, smell, taste and/or sound. Even when experiments were done by using "perception-enhancing instruments such as microscopes and telescopes," all data was ultimately *perceived qualitatively* through sense organs. However, soon scientists began to utilize quantity as a convenient method of describing the "eminently qualitative world" in which they lived. Categories like speed, angle and weight helped to differentiate the "relative differences between qualities." And from there, Kastrup argues that a shift occurred in which many scientists began ascribing the true nature of reality *only* to the quantitative aspect of their descriptions, which were determined through measurement. Because, unlike qualities, "only quantities can be objectively measured," many scientists advanced the theory that only quantities like charge or mass existed in the external world. Eventually, qualitative experiences became thought of as nothing more than the epiphenomena of neural activity. "This in a nutshell was the birth of metaphysical materialism," Kastrup concludes, "a philosophy that—absurdly—grants fundamental reality to mere *descriptions*, while denying the reality of which is described in the first place."[15] Scientific materialism, the belief that fundamental reality is constituted by matter that exists independently of consciousness, is what Kastrup calls *metaphysical materialism*. He argues materialism as such is a metaphysics because, while scientific experiments reveal how nature "*behaves*," the broader theory of materialism goes farther and makes a claim about what nature "essentially *is*."

Kastrup hypothesizes that the onset of materialism in the seventeenth century helped provide a sense of meaning at a time when the influence of religion on Western culture was diminishing. Jung also placed the rise of materialism in a similar historical context, although he traced the roots back even earlier. Jung located the pivot towards materialism in Europe as beginning during the Reformation, as the Gothic Age began to wane. He argued that the European mindset shifted from a vertical perspective (as represented by a "yearning for the heights" seen quite clearly in Gothic architecture), to a horizontal orientation, as evidenced by the beginning of modern sea voyaging, colonization and empirical discovery. In Jung's view, this meant that "consciousness ceased to grow upward, and grew instead in breadth of view, geographically as well as philosophically."[16] *Substantiality* became the hallmark of what was *real* and as time went on, substantiality was assigned more and more exclusively to material things, until at last, matter was declared the fundamental bedrock of reality.

Jung argued that the "horizontal perspective" that led to the exclusive privileging of matter was a direct reaction against the spiritual orientation and vertical perspective of the preceding era, and was in fact an *over*correction of the previous overemphasis on spirit (as in the nonmaterial). Jung argued that if we were "conscious of the *spirit of the age*," that is, the prevailing conscious and unconscious attitudes and tendencies of a given historical period, we would be able to recognize our fanatical subscription to materialism as an overcompensation for having once accounted for all physical phenomena in terms of spirit. He elucidated his point by saying,

> We delude ourselves with the thought that we know much more about matter than a 'metaphysical' mind or spirit, and so we overestimate material causation and believe that it alone affords us a true explanation of life. But matter is just as inscrutable as mind.[17]

If our consciousness had any temporal-historical continuity, Jung argued, then we would be able to recognize other moments in the history of Western thought when similar revolutions occurred, for instance, the fall of polytheism in ancient Greece. This might make us more critical of our philosophical assumptions, which are often mistaken for truth. Like Kastrup, Jung considered scientific materialism to be a "metaphysics of matter," one that thoroughly replaced a "metaphysics of mind" by the nineteenth century. Yet he saw both views as "equally logical, equally metaphysical, equally arbitrary, and equally symbolic."[18]

According to Jung, this kind of radical change in perspective and outlook could not have been made by "reasoned reflection" because "no chain of reasoning can prove or disprove the existence of either mind or matter." Both concepts are symbols that represent something unknowable. Considering, on one hand, consciousness to be a byproduct of material processes, and on the other that the "unpredictable behavior of electrons" is evidence of signs of "mental life even in them,"[19] are both equally speculative. However, as the spirit of the age shifted towards materialism, the previous unquestioning assumption that reality was created by the will of a "God who was a spirit" was usurped by the "equally unobservable truth that everything arises from material causes."[20] And so, by the nineteenth century, matter had been decreed as the ultimate seat of reality, and scientific thinkers and investigators "came to regard the mind as wholly dependent on matter and material causation."[21]

In his book *Decoding Jung's Metaphysics*, Kastrup points out that Jung anticipated by more than sixty years the "hard problem of consciousness." The hard problem of consciousness, a term coined by the philosopher David Chalmers in 1994, refers to the problem of explaining how and why physical processes in the brain give rise to subjective experiences of consciousness. Kastrup points to a line of Jung's from *Modern Man in Search of a Soul*—"we have literally no idea of the way in which what is psychic (i.e. experiential) can arise from physical elements, and yet cannot deny the reality of psychic events"—as evidence of such anticipation, but Kastrup himself offers an interesting and convincing rebuttal to the hard problem.

Kastrup believes that the "hard problem" is not something to be solved, but is instead something to be "seen through and circumvented." In Kastrup's analysis, beginning in the seventeenth century and culminating in the late nineteenth century, scientists ultimately forgot that quantity was brought into science as a way to discuss the differences between qualities. Subsequently, science replaced quality with quantity as the bedrock of fundamental reality. Kastrup argues that this move was like "cluelessly replacing reality with its description, the territory with the map"[22] and likens the change to "a painter who, having painted a self-portrait, points at it and proclaims himself to *be* the portrait." Matter, Kastrup argues, is a "metaphysical inference," a concept *created* by the mind, so when materialists attempt to "reduce mind to matter, they are effectively trying to reduce mind to one of mind's own conceptual creations."[23] Now that science has led us to believe that matter is the only thing that truly exists, and that consciousness is somehow a byproduct of material processes, we find ourselves in an impossible mental cul-de-sac. We invented quantity to discuss differences in quality, which is all we can ultimately know, and then we turned to matter, which we defined as something "purely quantitative" to see how it could produce the qualitative experience of consciousness. This makes the hard problem intractable and out of touch with reality, a metaphysical dilemma that Kastrup laments would be "comical if it weren't tragic."[24]

Kastrup argues that materialism renders matter more inaccessible and transcendent than "any ostensive spiritual world posited by the world's religion,"[25] because reality is experiential and we can experience spiritual worlds more than we can interact with matter as an objective entity that exists outside of our experience of it as a content of consciousness. Kastrup offers a succinct philosophical alternative to scientific materialism that he calls "analytical idealism," which he defines as "the notion that reality, while amenable to scientific inquiry, is fundamentally qualitative."[26] He says that because the world outside of individual experience *does* exist, there needs to be a way of thinking about the construction of this world at large. Materialism achieves this by placing matter in the position of the *real*, and our experiences of it are then explained as being various neuroscientific illusions. But Kastrup explains that the necessity of recognizing that there is something outside our individual minds is not the same as "having to posit something outside of *mind as a category.*" If we consider the world to be mental instead of material, and "postulate a *trans*personal field of mentation beyond our *personal* psyches," then there is no need to tie ourselves up in the Gordian knots of materialism. Because everything we know, we know qualitatively, then it is logical to extrapolate from this fact a larger transpersonal field of mentation. Our individual sets of mental qualities would then be modulated by a transpersonal set of qualities, a notion that Kastrup supports by suggesting that we witness this happening in our personal consciousnesses, when our qualitatively different emotions and thoughts "modulate each other all the time."[27]

Although there are differences, Kastrup's "transpersonal field of mentation" is not entirely unlike Jung's idea of the unconscious, which von Franz defines as a "psychic reality beyond ego consciousness."[28] Jung's ideas are consistent with Kastrup in that, under metaphysical materialism, spiritual experiences are more accessible than the transcendent (or trans-psychic) nature of matter. For Jung, psyche (as in the entire spectrum) is the primary material of reality, while the personal psyche (the middle portion of the spectrum) is that which experiences itself and the world. However, Jung would perhaps argue that spirit, as he defines it, and matter are equally unknowable in their "trans-psychic" essence—meaning in the essence beyond experience. Spirit, which might be defined in Jungian terms as the dynamic aspect of psyche, that which causes and sustains the perpetual movement of consciousness, is just as unknowable as matter, because while psyche is the ultimate substrate of reality, we can never truly know the mechanism that brings psyche to life and causes it to move.

Based on his own historical analysis of materialism, Jung would probably argue that privileging either matter or spirit as the fundamental seat of reality would be wrong. Too much emphasis in either direction necessitates a polarity that suggests a dualism, which then leads to issues like the hard problem of consciousness. Von Franz makes this plain when she writes that the separation between mind and matter, meaning the contrast we draw between our inner experiences versus that which we observe as being outside of ourselves, is only a subjective division, "only a limited polarization that our structure of consciousness imposes on us, but [one] that actually does not correspond to the wholeness of reality." She concludes her thought by saying, "it is rather to be suspected that these two poles (mental vs. material) actually constitute a unitary reality."[29]

Von Franz asserts that, in addition to underlying the images and stories seen in world mythologies and religions, archetypes "also underlie science in its intellectual premises."[30] What this means is that because the archetypes are *a priori* structuring elements within the collective unconscious, they organize the way we think and see both the inner world (as seen in religious and philosophical ideas and experiences) and the outer world (as seen in scientific formulas and theories). Because archetypes create patterns of themselves, and these patterns underlie both religious/philosophical and scientific conceptions, it makes sense that we would see an isomorphism between the patterns that appear in religious/philosophical and scientific dimensions. Von Franz thus draws a correlation between Heisenberg's uncertainty principle and the Greek philosopher Leucippus's theory that atoms had free will. She also highlights the similarities shared between Einstein/Minkowski's theory of space-time and both the concept of an omnipresent deity seen in many religions, as well as the stoic and neo-Platonic idea of the *pneuma*, the dynamic substance that constitutes the background of all reality (she quotes Plotinus's famous line, "God is an intelligible sphere whose center is everywhere and whose periphery is nowhere"[31]). She claims that "there is not a single important scientific paradigm that is not based on a primal archetypal intuition," and asserts that archetypal structures, which existed before personal ego consciousness, have "generated the themes of Western natural science." Matter, she writes, "is only one archetypal representation among many others" and is derived "from the archetype of the Great Mother."[32]

The archetype of the Great Mother, as defined by Jung, can be seen on both the personal and collective level in images like the actual mother, the grandmother, the goddess, the ancestress, the gnostic Sophia, Nature, Earth, the moon, the garden, the flower and the helpful animal. The Great Mother is whatever shelters, is fertile and fosters growth. In her negative aspects she is whatever seduces, devours and is hidden. The Father archetype, on the other hand, is seen in images like wind, spirits, demons, angels, the priest, the sage, the wizard, the wise professor and the helpful old man. The Father archetype is found in whatever is moving, inspiring and dynamic. Just as matter and the body corresponds to the Great Mother, spirit and the mind correspond to the Father archetype.

With regards to the concepts discussed in part one of this essay, there are direct correlations to be made between Jungian concepts and the ideas of quantum physics, as well as the individual theories put forth by the physicists themselves. When speaking about the nature of archetypes, von Franz writes that, while often distinct, archetypes are still "contaminated with one another. They merge into one another." She goes on to say that archetypes "are not like separate particles… but more like an 'electron smear.'" Although, as part one showed, the electron smear theory was eventually replaced with the theory of atomic orbitals—i.e. electron clouds—her point is still cogent. Von Franz also explains that archetypes are not simply separate entities that "swim around in the collective unconscious like pieces of bread in a soup," but are instead the "whole soup at every point and therefore always appears in specific mixtures."[33] This view is comparable to the analogy of the iceberg that the physicist Ruth Kastner uses to describe the relationship between the perceptible portion of reality and the quantum portion of reality. According to Kastner, the tip of the iceberg is what constitutes our everyday, *actualized* world, while the quantum portion of reality is analogized to the part of the iceberg that is submerged under the ocean, as well as the ocean itself. A further parallel with the archetypes is solidified in another aspect of the transactional interpretation of quantum mechanics, which Kastner champions; in that interpretation, the quantum world is that which organizes a collection of possibilities into the actuality of the spacetime dimension. Additionally, Kastner's argument that the quantum realms of reality, which exist outside of spacetime, are *real* even though they are not *actual* (as she defines the term) would be bolstered by Jung's attitude that "it is an almost ridiculous prejudice to assume that existence can only be physical."[34]

Another parallel can be drawn between Bohm's theory of the holographic universe/holoflux movement, Indra's net, and the collective unconscious. Bohm hypothesized that reality might have two separate levels. One level would be our everyday, tangible world, where Newtonian laws rule—he referred to this level as being of an "explicate order." And the other level, which underlies our own, would be a vast, potentially infinite level that births all physical objects—he referred to this level as being of an "implicate order." In the implicate order the concept of location disappears. The realm of implicate order is in essence a realm of pure information, and its informational nature means that it does not have to adhere to our three-dimensional concepts of space and time. Within the implicate order, two entangled particles are never actually separated at all, even though they might appear to be so in the explicate order. Instead, they exist in some kind of infinite informatic unity, and this in turn implies a certain wholeness of reality. In part one of this essay, Bohm's holographic universe was compared to Indra's Great Net of Gems, a mythological image of the universe as described in the *Avatamsaka Sutra*. On each node of Indra's great net, where its strands cross each other, there is a gem that contains and reflects all of the other gems in the net. Because the net is infinite, there are infinite gems, and each of the infinite gems that are reflected in a single gem in turn reflect all of the other gems in the net. Therefore, to look upon one is to look upon the whole.

There is a distinct parallel to be drawn between these concepts and myths and Jung's idea of the collective unconscious, which he described as not only an "identical psychic structure in all human beings," but also "an omnipresent continuum, an unextended-everywhere or an unextended presence." Jung meant that whenever something happens at a single point that touches or affects the collective unconscious, "it has happened *everywhere*." Von Franz argued that this explains the fact that the same idea is sometimes "discovered" at the same time "by two scientists working completely independent of each other."[35]

The physicist Hans-Peter Dürr believed that all of reality is made up of one single substance, and that our separateness from that substance is an illusion. However, for us to act in the world, we must break it up into pieces. Dürr defined the observer as a construction of consciousness that does just that: one who treats the world substance as something that can be divided and separated. Our universe is created as a whole in every second from some unobservable matrix. However, this process becomes smeared out at our human level of reality and because of this, the distinctions between matter and energy drawn in Newtonian laws of physics apply to us, even though our reality, at its base, is a unitary substance. For Dürr, what we call matter is just the result of us trying to make sense of our blurred-out position in the universe. Dürr called this kind of blurring, which also produces Cartesian dualism, "an infection."[36] Jung's ideas are consistent with nearly all of Dürr's conclusions regarding the unitary nature of reality, and both Jung and Kastrup's theories align well with Dürr's notion that Cartesian dualism is an infection.

There are also correlations to be drawn between Jungian formations and some scientific structures that were not mentioned in part one of this essay. Von Franz writes that our division of reality into material versus mental is "only a subjectively valid separation, only a limited polarization that our structure of consciousness imposes on us," but one that does not actually "correspond to the wholeness of reality."[37] This can easily be compared to the concept of *coarse-graining*, which is often invoked to explain why humans cannot absorb the information of the physical world in all its microscopic detail (and which is not unlike humanity's blurred-out position in the world as described by Dürr). Coarse-graining is used, in part, to discuss the nature of time, and the problem of why the human perception of time's linear flow is so different than time's apparent physical properties.

The passage of time is linked to the second law of thermodynamics because the second law of thermodynamics is the only law of physics that isn't *time symmetric*. Within all other equations that describe the rest of the laws of physics, a sequence of events can equally unfold backward or forward in time. However, the second law of thermodynamics *does* distinguish the future from the past—because of heat. The second law says that heat cannot, of itself, pass from a cold body to a hotter one. The physicist Carlo Rovelli emphasizes that "the arrow of time appears *only* when there is heat," which means that each and every time a difference between the past and the future materializes, heat is present.

Heat is the "microscopic agitation of molecules,"[38] and this "thermal agitation" creates disorder, also known as *entropy*. Rovelli explains that, somehow, through coarse-graining, we experience this disordering of molecules as the progression of time. However, if we were able to observe and process all the exact details of the microscopic level of reality, then what we deem to be the characteristic features of the progression of time would *vanish*. The neuroscientist Dean Buonomano offers a few excellent examples for thinking about this.

Imagine ten dice are thrown into a box and that box is shaken up. The dice will end up in a disordered placement and the system that comprises the box would be considered to be in a "state of high entropy."

However, if instead the dice were carefully arranged in a line and there was no shaking of the box, then the system would be highly ordered and in a "state of very low entropy."

To explain the relationship between entropy and the linear progression of time, Buonomano suggests we imagine that, instead of dice, two identical hydrogen atoms are placed on the left side of the box.

Later, one returns to the box and asks, "What is the most likely arrangement of the two atoms in terms of whether they are located on the left or right side of the box?" The atoms in the box may be in three possible configurations or states:

both atoms could be on the left side of the box (LL),

both could be on the right (RR),

or one atom could be on one side and one atom on the other.

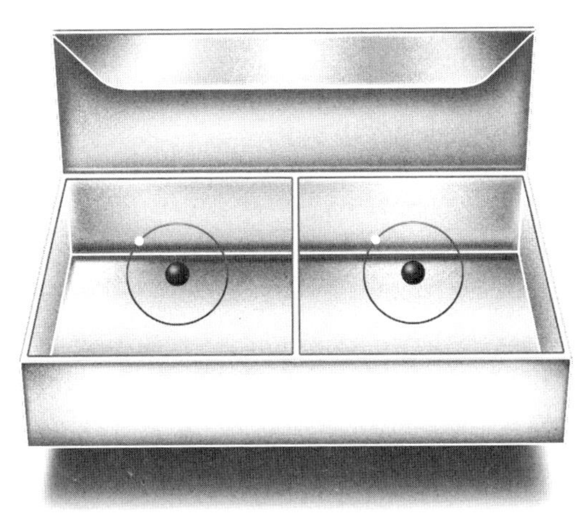

Because the "atoms are indistinguishable from each other," the states RL and LR would be "one and the same." So "the probability of each of these states, is ¼ LL, ¼ RR, and ½ LR(RL)." Considering that the RL and LR states are identical, because the atoms are identical, the box will most likely be found in a state in which the atoms are split evenly "because there are two ways to have one on each side."

If, after we have seen the atoms split evenly with one on each side, we close the box again and return later to check on it, "the chances that the state of the box will have "reversed" to its initial state [with both atoms on the left (LL)] is actually quite high." Now there would be a ¼ chance that the atoms will both be on the left, and the box will be in its original state.

Buonomano highlights the rather remarkable consequence of this by explaining that, if we were to think of the two atoms in the box as comprising "the entire universe, we might say that the universe went *back in time* because it returned to a state that is indistinguishable from its initial state."[iii, 39]

However, if we placed 10,000 hydrogen atoms on the left side of the box rather than just two,

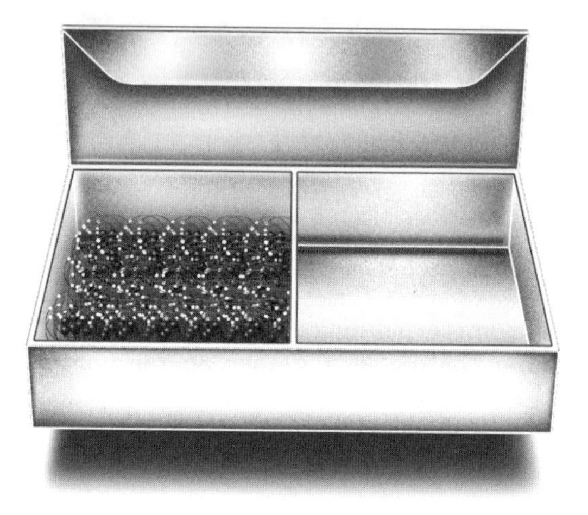

and "waited for the system to reach a state in which approximately half the atoms were on each side of the box,"

iii Emphasis added.

then the chances of the atoms reversing to a state where they are "all on the left is now inconceivably tiny."[iv] In this case, it becomes extremely improbable that the 10,000 atoms will ever return to their original state in the box. This conclusion is "very profound" because it can be interpreted "as meaning that the atoms in the box will not 'travel back' in time," and it is this exceptional unlikeliness that produces the arrow of time. This means that, rather than being a law in the same way that Newton's three laws of motion are, the second law of thermodynamics is instead a "statistical assertion that while reversing the state of an isolated system is ridiculously improbable, it is nevertheless legal."[40] So even though the second law of thermodynamics does not prohibit such reversal of states (for instance a broken glass returning to an unbroken state), it does render the situation virtually impossible.

If one could observe reality at the microscopic level, the difference between the past and the future would disappear. Something in our consciousness converts fine-grained microscopic systems into a coarse-grained understanding in which some of the detail of the system is blurred out. And this can be directly compared to von Franz's view that the separation between matter and mind results from "a limited polarization that our structure of consciousness *imposes on us*, but that actually does not correspond to the wholeness of reality." In both cases our consciousness performs a process that does not allow us to see the entirety of reality for what it really is.

Likewise, there are other religious ideas that seem to align well with Jungian concepts. In Jung's light spectrum analogy for psyche, subjective freedom lies only in the very middle of the spectrum. If an individual ego moves too far toward either end, then that ego falls under the influence or possession of either the instinctual/somatic pole or the archetypal pole. This seems to parallel the concept of the Middle Way seen in Buddhist traditions. The Middle Way, which was one of the first teachings that Buddha gave his followers, teaches that enlightenment can be achieved by walking the middle path between indulgence and asceticism. According to the Buddha, both deprivation and indulgence were detrimental to the quest for Nirvana, and freedom to progress towards awakening lie only in the middle of such extremes.

iv "much, much smaller than one in a googol" (Buonomano, 152)

In considering the correspondences between Eastern religious philosophies and the worldview ushered in by quantum mechanics—a confluence of fields that is referred to as "quantum mysticism"—the organizational function of the archetypes provides an interesting theoretical interpretation. The explanation often given for the correlations found between quantum physics and Eastern religions is that modern science has rediscovered/is rediscovering ancient wisdom. Many modern scientists balk at this notion, contending that the correlation distorts the formalism of quantum mechanics. However, if the issue is seen through a Jungian lens, then the "rediscovery" explanation can be circumvented. If archetypes are responsible for organizing both religious and mythological symbols and concepts on the one hand, and scientific theories and ideas on the one other, then the isomorphism seen in the theories of quantum physics and the concepts of Eastern religions can be explained by the fact that both arise from the same archetype. Some archetypal formation, originating from and organizing itself within the collective unconscious, becomes expressed in two different ways, through science and through religion.

These correspondences point to a shared origin in the organizational tendencies of psyche as it experiences and describes reality. However, Jung not only viewed the archetypes as creating scientific concepts and religious ideas—both of which, one might argue, are mental formations—but he also suggested that physical Nature *itself* is organized via archetypes. Kastrup elucidates this point beautifully in *Decoding Jung's Metaphysics*. He provides a strong argument for his interpretation of Jung's *monism*, which here is taken to mean a philosophical position that refutes the fundamental distinction between mind and matter. Kastrup quotes a passage from one of Jung's letters,

> The collective unconscious is identical with Nature to the extent that Nature herself, including matter, is unknown to us. I have nothing against the assumption that the psyche is a quality of matter or matter the concrete aspect of the psyche, provided that "psyche" is defined as the collective unconscious.[41]

Kastrup interprets Jung's metaphysical views, as demonstrated in the above quote, as being that psyche (as the collective unconscious) and Nature (as matter) are in fact *one and the same thing*. This one and the same thing "impinges" on our ego-consciousness, which is all we can ever know, through "the sense organs (in the case of the objective world) or through a shared, internal psychic boundary (in the case of the objective psyche)."[42] Kastrup points out the "symmetry and elegance" of Jung's vision. The external world "as it is in itself" is autonomous and transcendent and abuts on "ego-consciousness through the sense organs, generating the autonomous imagery of perception." The collective unconscious "as it is in itself" is similarly autonomous and transcendent, and abuts on "ego-consciousness through a shared internal psychic boundary (the collective unconscious), generating the autonomous imagery of dreams and visions."[43] Both spheres of reality are real because they "act" upon us, creating the experience of life itself, even though their trans-psychic essence can never be known. Kastrup's analysis is illustrated by a version of this diagram:

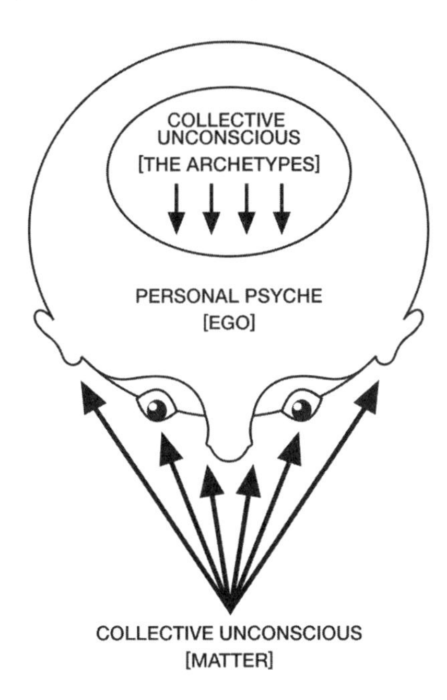

The equivalence between Nature and psyche (as the collective unconscious) means that, along with the contents of consciousness, the outer world—matter itself—is organized by archetypes. This goes beyond von Franz's exposition of the role archetypes play in structuring scientific ideas and formulas, which may be viewed as mental frameworks or theories about the outer world. What Kastrup is saying is that archetypes organize *the outer world itself*. Kastrup points to Jung's concept of *synchronicity* to bolster his interpretation. Synchronicity, as Jung defined it, involves circumstances or events that share some kind of common symbolism, those that relate to each other through meaning but not necessarily through material causal connection. Most often synchronicities involve some content of consciousness aligning meaningfully with an event that happens in the external world. Kastrup argues that, because the archetypes within the collective unconscious order the contents that emerge into consciousness, and because the collective unconscious and Nature are one and the same, the archetypes must order Nature too. When the archetypal structures line up, there we have synchronicities.

Extended beyond the classical definition of synchronicity, it is possible to apply this concept to the structure of the quantum-entangled photons imaged by researchers from the Sapienza University of Rome and the University of Ottawa in Canada, the uncanniness of which was discussed in part one of this essay.

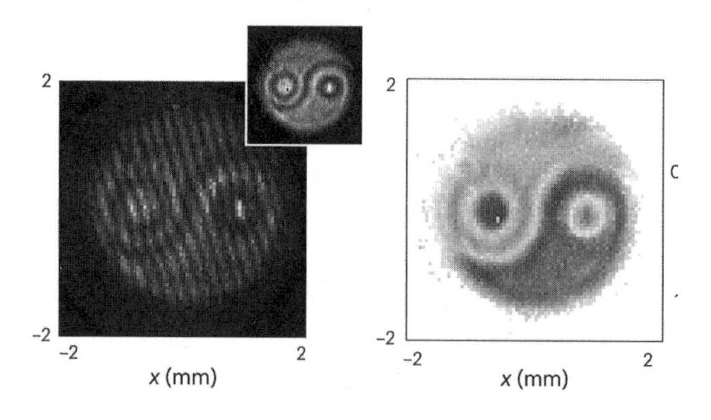

(Image credit: *Nature Photonics*, Zia et al.)

The fact that the image of entangled photons makes a perfect yin-yang cannot be avoided. Rather than a remarkable coincidence, perhaps it is a result of the archetypal structuring patterns that organize both consciousness *and* the physical world.

Von Franz had a slightly different formulation as she interpreted Jung's overall metaphysics. She presented Jung's spectrum analogy with a dotted line underneath that connects the infrared and ultraviolet poles of psyche.[v]

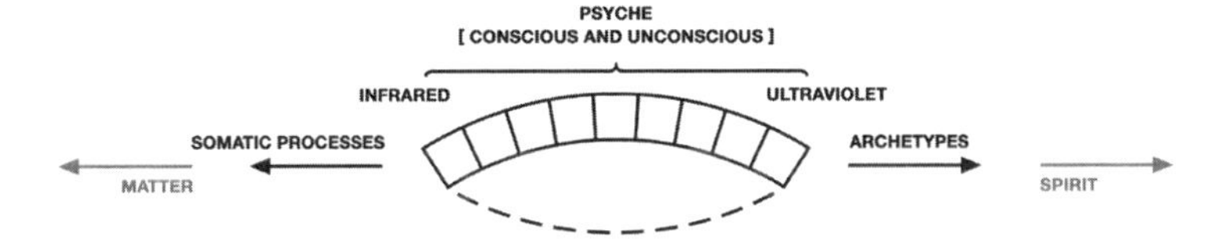

For von Franz, the dotted line represents "the unitary reality of psyche (as a whole) and matter," but cautioned that "this sphere cannot however be observed directly."[44] Perhaps a better image of the unitary reality put forth by Jung and elucidated by von Franz would be this:

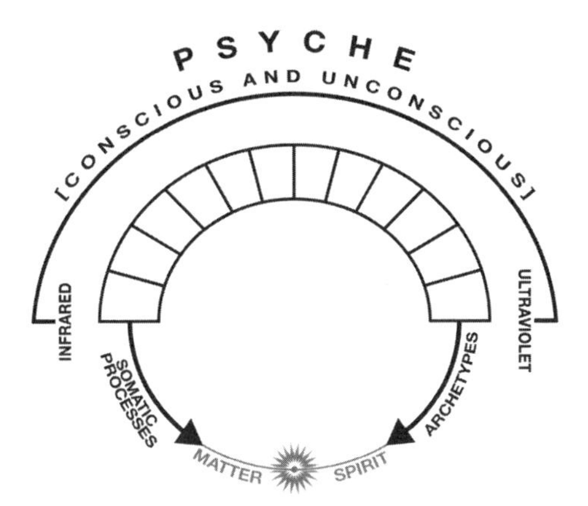

As a two-dimensional image, the bottom of the circle represents the point where matter and psyche meet and mix together. However, if we think of this image as representative of a three-dimensional

v Note: in the original diagram of the spectrum used by von Franz, she did not include arrows extending to "matter" and "spirit," those were the addition of this author.

"sphere" as von Franz suggests, then there would be many points throughout the sphere where matter and psyche mix together in ever-changing ratios. In fact, it could be surmised that at each and every point where there is matter, *there is also some aspect of psyche.* Similarly, there would be an aspect of the material throughout the entirety of psyche, and therefore some aspect of matter in consciousness (von Franz argued a version of this idea with regards to paranormal activity, such as the appearance of a ghost). And although von Franz cautioned that we cannot "directly" observe this portion of the sphere, perhaps we can at times observe it indirectly. There must be places where matter and psyche meet, and still other places where they transform into each other—or rather into ratios where one element is dominant enough so that the presence of the other is hidden. Perhaps this is what is seen in quantum phenomena like the instantaneous collapse of an electron's wave function upon measurement, or in other instances of what is called the *observer effect.*

The most well-known example of the observer effect in quantum physics evolved from the electron gun double-slit experiment covered in part one of this essay. The set-up in the later version of the experiment is similar to the original experiment, with an electron gun that shoots beams of identical electrons positioned in front of one screen with two slits in it. Behind that screen, there is a second screen capable of recording where an electron has hit it by registering the electron's position with a little bright dot. The only difference from the original experiment is that there is now a "detector" placed behind the first screen that detects which slit each electron passes through on its way to the back screen.

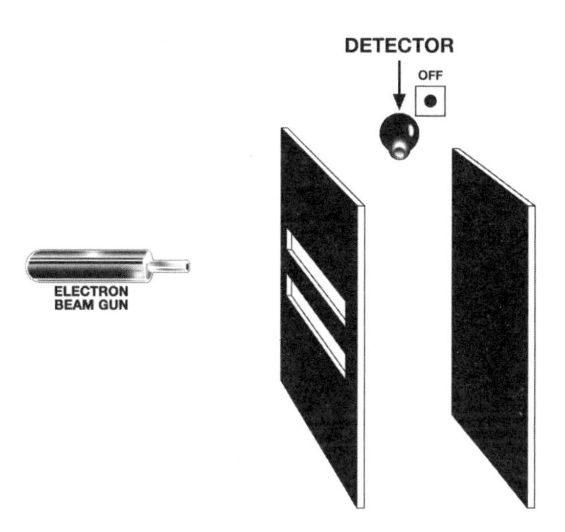

As previously discussed, each electron leaves the gun as a localized particle and is recorded on the back screen as a localized particle, but the multiple electron strikes on the back screen exhibit the pattern of a wave that has spread out, split in two, and interfered with itself. Even if the electrons from the gun are sent one by one, the back screen eventually registers an interference pattern, the same kind that are seen in waves. However, something incredibly strange happens once you add the presence of the detector. Once the detector is turned on, *suddenly the interference pattern disappears*.

If electrons from an electron gun are sent through the two slits in the presence of a switched-on detector, they collect on the backscreen in the form of the slits they pass through, just as we would expect them to from the perspective of classical (Newtonian) physics.

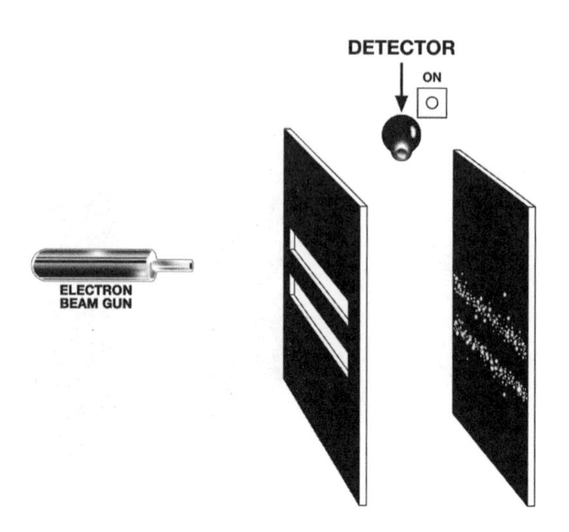

Yet once the detector is turned off again, the interference pattern reemerges.

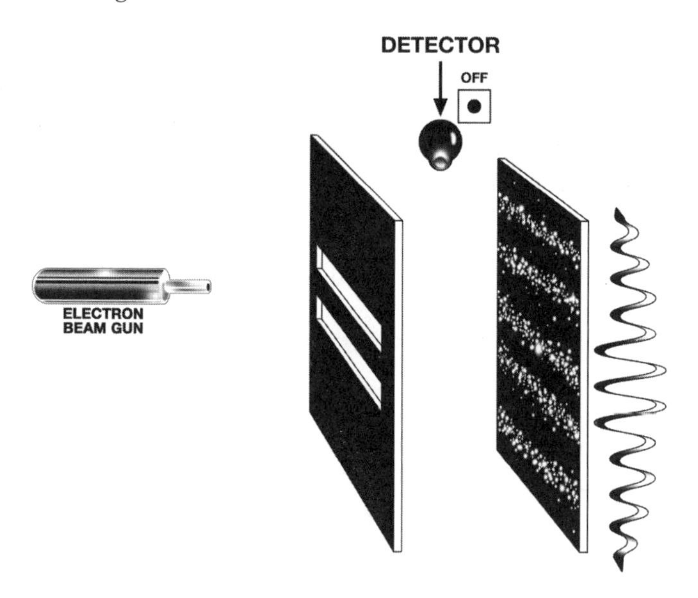

Somehow the electrons behave differently when they are being observed versus when they are not being observed. Only when being observed do the electrons behave as a particle throughout the entirety of the experiment. When they are not being observed, they leave the electron gun as a particle, arrive at the back screen as a particle, and behave as a wave in between. Quantum physics concludes that the act of observing something changes that which is observed. But how can we explain the fact that the electrons seem to know when they are being observed?

From a Jungian perspective, perhaps this phenomenon can be thought of as happening at a point where *matter and psyche meet*. Each point in reality have may have a different ratio of matter to psyche, and the points may be distributed in different patterns throughout reality's unitary substance. The quantum observer effect may show that if we dig deeply enough into the nature of matter, we are suddenly confronted with consciousness. There is another concept Jung employs that may help us to understand this, called *enantiodromia*. Jung uses enantiodromia to describe "the emergence of the unconscious opposite in the course of time,"[45] meaning that any extreme tendency in a psychic system will one day become its opposite. This occurs when a one-sided position or tendency has come to dominate one's life. Because Nature strives towards maintaining balance, a countertendency appears to reinstate equilibrium and the sudden inversion is the enantiodromia. Enantiodromia does not just apply to an individual. Jung extended his use of the term out to social constructions, writing that "the rational attitude of culture necessarily runs into its opposite, namely the irrational devastation of culture."[46] He actually borrowed the term from Heraclitus, who used it as a means of explaining the way that everything in Nature eventually runs into its opposite ("Cold warms up, warm cools off, moist parches, dry dampens"[47]). The word itself breaks down into the roots *enantios*, meaning "opposite," and *dromos*, meaning "running."

Extending this principle toward the nature of reality itself and the division of matter and psyche, we can see the limits of our consciousness as it collides with matter—just as Heisenberg's uncertainty principle says that we can never know both a particle's position and its speed/momentum at the same time. Similarly, in the quantum observer effect, we can see the limits of matter as it collides with our consciousness.

Perhaps these are instances of Nature maintaining her balance, so that as we inquire deeper and deeper into the nature of our physical reality, we suddenly see matter display something like consciousness. We probe the very stitches that fasten the fabric of our material world, only to see the stitches dancing before our eyes. It is as though we have stuck our hand through a mirror, only then to feel ourselves tickling the back of our own neck. Or, in the words of astronomer, physicist and mathematician Sir Arthur Eddington,

> We have found that where science has progressed the farthest, the mind has but regained from nature that which the mind has put into nature. We have found a strange footprint on the shores of the unknown. We have devised profound theories, one after another, to account for its origin. At last, we have succeeded in reconstructing the creature that made the footprint. And lo! it is our own.[48]

Matter is myth. It is a myth that keeps us tethered to a world that has, at its core, a nonmaterial question mark. Reality is a psycho-physical substance that has no beginning or end. Mind and matter only exist because of each other, or more accurately because they *are* each other—and the apparent border that separates them is fluid and ever-changing. In the dissolution of the boundaries of materialism there is liberation, a freedom to accept things as they really are. Everything you see is inside of you and you are inside of all that you see. In this exchange, there is no primary substance—only a dance between two expressions of one infinite mystery.

NOTES

A Letter to the Reader

1 Jeffrey J. Kripal and Whitley Strieber, *The Super Natural: A New Vision of the Unexplained* (Penguin Random House, 2016), 113.

Pluto & the Mythic Dimension

1 James Hillman, *The Dream and the Underworld* (William Morrow, 1979), 28.

2 James Hillman, *A Blue Fire* (Harper Perennial, 1989), 15–16.

3 Hillman, *The Dream and the Underworld*, 29.

4 J.F. Martel, "Pluto and the Death of God," in *Pluto: New Horizons for a Lost Horizon*, ed. Richard Grossinger (North Atlantic Books, 2015), 160–161.

5 Paxton Scott and William Yin, "Bill Gates, Gov. Gavin Newsom speak at unveiling of new human-centered artificial intelligence institute," *The Stanford Daily*, March 19, 2019, https://stanforddaily.com/2019/03/19/bill-gates-gov-gavin-newsom-speak-at-unveiling-of-new-human-centered-artificial-intelligence-institute.

6 Geoffrey Hinton, "The so-called 'Godfather of the A.I.' joins The Lead to offer a dire warning about the dangers of artificial intelligence," Interview by Jake Tapper, *The Lead with Jake Tapper,* CNN, May 2, 2023. Video, 4:27, https://www.cnn.com/videos/tv/2023/05/02/the-lead-geoffrey-hinton.cnn.

7 Hinton, interview.

8 Ralph E. Lapp, "The Einstein Letter That Started It All," *The New York Times,* August 2, 1964, https://www.nytimes.com/1964/08/02/archives/the-einstein-letter-that-started-it-all-a-message-to-president.html.

9 Kai Bird and Martin J. Sherwin, *American Prometheus: The Triumph and Tragedy of J. Robert Oppenheimer* (A.A. Knopf, 2005), 332.

10 "Pause Giant AI Experiments: An Open Letter," Future of Life Institute, published March 22, 2023, https://futureoflife.org/open-letter/pause-giant-ai-experiments/.

11 Eliezer Yudkowsky, "Pausing AI Developments Isn't Enough. We Need to Shut it All Down," *Time*, March 29, 2023, https://time.com/6266923/ai-eliezer-yudkowsky-open-letter-not-enough/.

12 "Statement on AI Risk," Center for AI Safety, published May 30, 2023, https://www.safe.ai/work/statement-on-ai-risk.

13 Cade Metz, "The ChatGPT King Isn't Worried, but He Knows You Might Be," *The New York Times*, March 31, 2023, https://www.nytimes.com/2023/03/31/technology/sam-altman-open-ai-chatgpt.html.

14 Pearl S. Buck, "The bomb—the end of the world?," *The American Weekly*, March 8, 1959, http://large.stanford.edu/courses/2015/ph241/chung1/docs/buck.pdf.

15 Sam Altman, Greg Brockman and Ilya Sutskever, "Governance of superintelligence," Open AI, May 22, 2023, https://openai.com/index/governance-of-superintelligence/.

16 Ian Prasad Philbrick and Tom Wright-Piersanti, "A.I. or Nuclear Weapons: Can You Tell These Quotes Apart?," *The New York Times*, June 10, 2023, https://www.nytimes.com/2023/06/10/upshot/artificial-intelligence-nuclear-weapons-quiz.html.

17 Buck, "The bomb—the end of the world?"

18 Hillman, *The Dream and the Underworld*, 28.

19 Giorgio Agamben and Monica Ferrando, *The Unspeakable Girl: The Myth and Mystery of Kore* (Seagull Books, 2014), 6.

20 Saffron Rossi, *The Kore Goddess: A Mythology & Psychology* (Winter Press, 2021), 29.

21 Rossi, *The Kore Goddess*, 133.

22 Rossi, *The Kore Goddess*, 64.

23 Rossi, *The Kore Goddess*, 64.

24 Rossi, *The Kore Goddess*, 63.

Photograph taken by Daniel Villafruela, cropped and made black & white. License may be found at https://commons.wikimedia.org/wiki/File:Ch%C3%A2teau_de_Chantilly-Propserpine_(Chapu)-20120917.jpg.

Myths + Models of Time & Timelessness

1 Carlo Rovelli, *The Order of Time* (Riverhead Books, 2017), 3.

2 Dean Buonomano, *Your Brain is a Time Machine* (W.W. Norton, 2017), 11.

3 Rovelli, 12.

4 Rovelli, 16.

5 Brian Greene, host, NOVA, season 38, episode 17, "The Fabric of the Cosmos: The Illusion of Time," PBS, November 8, 2011, 53 min., 40 sec., https://www.pbs.org/video/nova-the-fabric-of-the-cosmos-the-illusion-of-time.

6 Hermann Minkowski, "Space and Time," lecture delivered before the Naturforscher Versammlung (Congress of Natural Philosophers) at Cologne, September 21, 1908.

7 Jonathan Schooler, interview by LD Deutsch, March 9, 2022.

8 Buonomano, 216.

9 Buonomano, 13.

10 Buonomano, 6.

11 Buonomano, 178.

12 The profoundness of this idea can perhaps be seen in the pervasiveness of relativity that is omnipresent throughout the universe and warps so completely the fabric of spacetime.

13 Harold Kelman, "Kairos: The Auspicious Moment," *The American Journal of Psychoanalysis*, 29, 59–83 (1969), 80 https://doi.org/10.1007/BF01872669.

14 Kelman, 67.

15 Marie-Louise von Franz, *Time: Rhythm and Repose* (Thames and Hudson, 1978), 9.

16 von Franz, *Time: Rhythm and Repose*, 7.

17 Marie-Louise von Franz, *On Divination and Synchronicity: The Psychology of Meaningful Chance* (Inner City Books, 1980), 8.

18 von Franz, *On Divination*, 13.

19 Frank Swetz, *Legacy of the Luoshu: The 4,000 Year Search for the Meaning of the Magic Square of the Order Three* (A.K. Peters/CRC Press, 2008), 21.

20 Swetz, 23.

21 Marie-Louise von Franz, *Number and Time* (Northwestern University Press, 1974), 237.

22 von Franz, *Time: Rhythm and Repose*, 19.

23 C.G. Jung, *Dream Analysis 1: Notes on the Seminar Given in 1928–1930* (Princeton University Press, 1984), 417.

24 John Archibald Wheeler, "American Physicist," *Scientific American*, Vol. 267 (1992).

25 Rovelli, 125.

Photographs:

Kairos:
Kairos, relief (Museo di Antichità, Turin). License may be found at https://www.researchgate.net/figure/Kairos-relief-Museo-di-Antichita-Turin-Sl-5-Kairos-relief-Museo-di-Antichita_fig5_372376281

Chronos:
"Time," statue by Franz Ignaz Gunther, Germany, c. 1765–70, Bayerisches Nationalmuseum, Munich

Aion 1:
Relief with Aion-Phanes, 125-50 AD, Galleria Estense, Modena, Italy. License may be found at https://en.wikipedia.org/wiki/Galleria_Estense#/media/File:Arte_romana,_rilievo_con_aion-phanes_entro_lo_zodiaco,_150_dc_ca.,_probabilmente_da_un_mitreo.jpg

Aion 2:
Leontocephaline from Villa Albani, white marble statue, H. 155 cm, Br. 37 cm (base), Vatican Museums, Rome (190–225 C.E.)

Technomythology

1 Marie-Louise von Franz, *Creation Myths, Revised Edition* (Shambhala Publications, 1995), 1.

2 Marie-Louise von Franz, *Psyche & Matter* (Shambhala Publications, 2001), 6.

3 von Franz, *Creation Myths*, 2.

4 von Franz, *Creation Myths*, 9.

5 von Franz, *Creation Myths*, 16.

6 David Streitfeld, "Silicon Valley Confronts the Idea That the 'Singularity' Is Here," *The New York Times*, June 11, 2023, https://www.nytimes.com/2023/06/11/technology/silicon-valley-confronts-the-idea-that-the-singularity-is-here.html.

7 von Franz, *Creation Myths*, 11.

8 See LD Deutsch, "Myths + Models of Time & Timelessness"

9 Joshua Rothman, "What Are the Odds We Are Living in a Computer Simulation?" *The New Yorker*, June 9th, 2016, https://www.newyorker.com/books/joshua-rothman/what-are-the-odds-we-are-living-in-a-computer-simulation#:~:text=Citing%20the%20speed%20with%20which,just%20%E2%80%9Cone%20in%20billions.%E2%80%9D.

10 Jacques Vallée, "A Theory of Everything (else)" *TEDxBrussels* YouTube Channel, published November 23, 2011, https://www.youtube.com/watch?v=S9pR0gfil_0&lc=UgicUo6CCswWtXgCoAEC.

11 Vallée, "A Theory of Everything (else)"

12 Vallée, "A Theory of Everything (else)"

13 Erik Davis, *TechGnosis: Myth, Magic & Mysticism in the Age of Information* (North Atlantic Books, 2015), 5.

14 John D. Barrow, *New Theories of Everything: The Question for Ultimate Explanation* (Oxford University Press, 2007), 11.

15 Kripal and Strieber, The Super Natural, 19.

16 C.G. Jung, "The Gifted Child," *The Collected Works of C.G. Jung, Vol. 17* (Princeton University Press, 1981) 133–145.

17 Mark Whitney, *Matter of Heart*, Kino International, 1986, 1hr, 53min https://www.youtube.com/watch?v=Ed3vPb9bmcw.

18 Edward F. Edinger, *Archetype of the Apocalypse: A Jungian Study of the Book of Revelation* (Open Court, 2002), 3.

19 von Franz, *Creation Myths*, 11.

The Myth of Matter, Part I

1 Robert Boyle, *The Sceptical Chymist or, Chymico-physical doubts & paradoxes*, printed by J. Cadwell for J. Crooke, 1661.

2 Isaac Newton, *Philosophiae Naturalis Principia Mathematica*, 1687.

3 Ruth Kastner, *Understanding Our Unseen Reality: Solving Quantum Riddles* (Imperial College Press, 2015), 26.

4 Ruth Kastner, *The Transactional Interpretation of Quantum Mechanics: The Reality of Possibility* (Cambridge University Press, 2013), 2.

5 Werner Heisenberg, *Physics and Philosophy* (Harper & Row, 1962), xii.

6 Erwin Schrödinger, *My View of the World* (Cambridge University Press, 1964), 19.

7 Kastner, *The Transactional Interpretation of Quantum Mechanics*, 74.

8 Kastner, *The Transactional Interpretation of Quantum Mechanics*, 75.

9 John A. Wheeler, "A Call on Physicist John A. Wheeler," *University: A Princeton Quarterly*, Summer, 1972.

10 David Bohm: as quoted in Jane Carroll's "Wholeness, Timelessness and Unfolding Meaning," *Beshara Magazine*, Issue 14, Winter 2020.

11 Erwin Schrödinger, *What is Life & Mind and Matter* (Cambridge University Press, 1967), 130.

12 Fritjof Capra, *Uncommon Wisdom* (Simon & Schuster, 1988), 43.

13 Fritjof Capra, "Heisenberg and Tagore," *Fritjof Capra*, July 3, 2017, https://www.fritjofcapra.net/heisenberg-and-tagore.

14 Niels Bohr, from a speech given on quantum theory at Celebrazione Secondo Centenario della Nascita di Luigi Galvani, Bologna, Italy, October, 1937.

15 Ruth Kastner, *Adventures in Quantumland: Exploring Our Unseen Reality* (WSPC Europe, 2019), 18.

16 Kastner, *Adventures in Quantumland*, 17.

17 Ruth Kastner, interview by David Garofalo, "The Quantum Handshake Welcoming You to Spacetime," *David Garofolo's Corner: From Science to Culture, Black Holes to the Multiverse*, December, 24, 2020, https://www.davidgarofaloscorner.com/post/the-quantum-handshake-welcoming-you-to-spacetime-with-physicist-and-philosopher-dr-ruth-kastner.

18 Max Planck, from a lecture entitled "Das Wesen der Materie" [The Essence/Nature/Character of Matter], given in Florence, Italy, 1944.

19 Hans-Peter Dürr, interview by Deepak Chopra, "Quantum Physics & Creativity—Hans-Peter Dürr and Deepak Chopra," *QuantenPhysik* YouTube Channel, published March 20, 2014, https://www.youtube.com/watch?v=sIKgldFGPe0.

20 Rovelli, *The Order of Time*, 98–99.

21 Albert Einstein: as quoted by Aylesa Forsee in *Albert Einstein, Theoretical Physicist* (MacMillan, 1963), 81.

The Myth of Matter, Part II

1 C.G. Jung, *Psychological Types* (Princeton University Press, 1971), 463.

2 von Franz, *Psyche & Matter*, 2.

3 Language adapted from paragraph: "Everything of which I know, but of which I am not at the moment thinking; everything of which I was once conscious but have now forgotten; everything perceived by my senses, but not noted by my conscious mind; everything which, involuntarily and without paying attention to it, I feel, think, remember, want, and do; all the future things which are taking shape in me and will sometime come to consciousness; all this is the content of the unconscious," in C.G. Jung, *The Structure and Dynamics of the Psyche* (Princeton University Press, 1970), 185.

4 von Franz, *Psyche & Matter*, 3.

5 C.G. Jung, *The Archetypes and the Collective Unconscious* (Princeton University Press, 1969), 42.

6 Bernardo Kastrup, *Decoding Jung's Metaphysics: The Archetypal Semantics of an Experiential Universe* (Iff Books, 2021), 34.

7 von Franz, *Psyche & Matter*, 6.

8 von Franz, *Psyche & Matter*, 28.

9 von Franz, *Psyche & Matter*, 10.

10 von Franz, *Psyche & Matter*, 157.

11 von Franz, *Psyche & Matter*, 5.

12 Kastrup, *Decoding Jung's Metaphysics*, 85.

13 von Franz, *Psyche & Matter*, 157.

14 von Franz, *Psyche & Matter*, 16.

15 Bernardo Kastrup, *Science Ideated: The Fall of Matter and the Contours of the Next Mainstream Scientific Worldview* (Iff Books, 2021), 1.

16 Jung, *The Structure and Dynamics of the Psyche*, 339.

17 Jung, *The Structure and Dynamics of the Psyche*, 342.

18 Jung, *The Structure and Dynamics of the Psyche*, 340.

19 Jung, *The Structure and Dynamics of the Psyche*, 339.

20 Jung, *The Structure and Dynamics of the Psyche*, 8, 340.

21 Jung, *The Structure and Dynamics of the Psyche*, 8, 339.

22 Kastrup, *Science Ideated*, 2.

23 Kastrup, *Science Ideated*, 10.

24 Kastrup, *Science Ideated*, 2.

25 Kastrup, *Science Ideated*, 10.

26 Kastrup, *Science Ideated*, 3.

27 Kastrup, *Science Ideated*, 10.

28 von Franz, *Psyche & Matter*, 2.

29 von Franz, *Psyche & Matter*, 11.

30 von Franz, *Psyche & Matter*, 11.

31 von Franz, *Psyche & Matter*, 12.

32 von Franz, *Psyche & Matter*, 15.

33 von Franz, *Psyche & Matter*, 9.

34 Jung, *Psychology and Religion: East and West*, (Princeton University Press, 1969), 12.

35 von Franz, *Psyche & Matter*, 157.

36 Dürr, interview.

37 von Franz, *Psyche & Matter*, 11.

38 Rovelli, 28.

39 Buonomano, 152.

40 Buonomano, 153.

41 C.G. Jung, *Letters: Vol. 2.*, (Princeton University Press, 1976), 540.

42 Kastrup, *Decoding Jung's Metaphysics*, 94.

43 Kastrup, *Decoding Jung's Metaphysics*, 95.

44 von Franz, *Psyche & Matter*, 11.

45 C.G. Jung, *Psychological Types*, 426.

46 C.G. Jung, *Two Essays on Analytical Psychology*, (Princeton University Press, 1972), 72.

47 Heraclitus, *The Art and Thought of Heraclitus: An Edition of the Fragments with Translation and Commentary*, ed. Charles H. Kahn (Cambridge University Press, 1981), Fragment XLIX.

48 Arthur S. Eddington, *Space, Time and Gravitation: An Outline of the General Relativity Theory* (Cambridge University Press, 1920), 200.

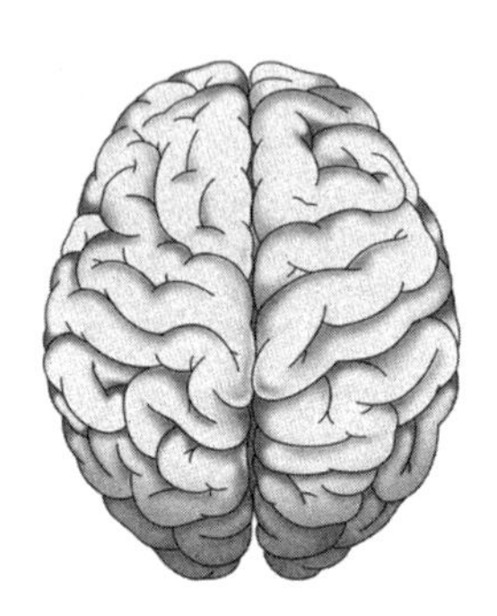

BIBLIOGRAPHY

Agamben, Giorgio, and Monica Ferrando. *The Unspeakable Girl: The Myth and Mystery of Kore*. Seagull Books, 2014.

Al Khalili, Jim. *Quantum: A Guide for the Perplexed*. Weidenfeld & Nicolson, 2012.

Altman, Sam, Brockman, Greg, and Sutskever, Ilya. "Governance of superintelligence." OpenAI, May 22, 2023.
https://openai.com/index/governance-of-superintelligence/

Bailenson, Jeremy, and Blascovich, Jim. *Infinite Reality*. HarperCollins, 2011.

Barrow, John D. *New Theories of Everything: The Question for Ultimate Explanation*. Oxford University Press, 2007.

Bird, Kai, and Sherwin, Martin J. *American Prometheus: The Triumph and Tragedy of J. Robert Oppenheimer*. A.A. Knopf, 2005.

Brown, Mike. *How I Killed Pluto and Why It Had It Coming*. Spiegel & Grau, 2012.

Bohr, Niels. From a speech given on quantum theory at Celebrazione Secondo Centenario della Nascita di Luigi Galvani. Bologna, Italy. October, 1937.

Boyle, Robert. *The Sceptical Chymist or, Chymico-physical doubts & paradoxes*. Printed by J. Cadwell for J. Crooke, 1661.

Buck, Pearl S. "The bomb — the end of the world?" *The American Weekly*, March 8, 1959.
http://large.stanford.edu/courses/2015/ph241/chung1/docs/buck.pdf

Buonomano, Dean. *Your Brain is a Time Machine. W.W. Norton, 2017*

Campbell, Joseph, ed. *Man and Time: Papers from the Eranos Yearbook*. Princeton University Press, 1957.

Capra, Fritjof. "Heisenberg and Tagore." *Fritjof Capra Blog*. July, 3 2017.
https://www.fritjofcapra.net/heisenberg-and-tagore

Capra, Fritjof. *Uncommon Wisdom*. Simon & Schuster, 1988.

Carrol, Jane. "Wholeness, Timelessness and Unfolding Meaning." *Beshara Magazine*, Issue 14, Winter 2020.

Center for AI Safety. "Statement on AI Risk." Published May 30, 2023.
https://www.safe.ai/work/statement-on-ai-risk

Cook, Francis H. *Hua-yen Buddhism: The Jewel Net of Indra*. The Pennsylvania State University Press, 1977.

Davis, Erik. *TechGnosis: Myth, Magic & Mysticism in the Age of Information*. North Atlantic Books, 2015.

Dürr, Hans-Peter. Interview by Deepak Chopra. "Quantum Physics & Creativity — Hans-Peter Dürr and Deepak Chopra." *QuantenPhysik* YouTube Channel. Published March 20, 2014.
https://www.youtube.com/watch?v=sIKgldFGPe0

Dyson, George. *Turing's Cathedral: The Origins of the Digital Universe*. Knopf Doubleday, 2012.

Eddington, Arthur S. *Space, Time and Gravitation: An Outline of the General Relativity Theory*. Cambridge University Press, 1920.

Farman, Abou. *On Not Dying: Secular Immortality in the Age of Technoscience*. University of Minnesota Press, 2020.

Edinger, Edward F. *Archetype of the Apocalypse: A Jungian Study of the Book of Revelation*. Open Court, 2002.

Forsee, Aylesa. *Albert Einstein, Theoretical Physicist*. MacMillan, 1963.

Fraser, J.T. *Of Time, Passion and Knowledge*. Princeton University Press, 1975.

Future of Life Institute. "Pause Giant AI Experiments: An Open Letter." Published March 22, 2023.
https://futureoflife.org/open-letter/pause-giant-ai-experiments/

Gieser, Suzanne. *The Innermost Kernal: Depth Psychology and Quantum Physics: Wolfgang Pauli's Dialogue with C.G. Jung*. Springer Berlin, 2005.

Greene, Brian, host. *NOVA*. Season 38, episode 17. "The Fabric of the Cosmos: The Illusion of Time," PBS, November 8, 2011. 53 min., 40 sec.
https://www.pbs.org/video/nova-the-fabric-of-the-cosmos-the-illusion-of-time

Grimal, Pierre. *The Dictionary of Classical Mythology*. Blackwell, 1986.

Hayles, N. Katherine. *How We Became Posthuman*. University of Chicago, 1999.

Hayles, N. Katherine. *How We Think: Digital Media and Contemporary Technogenesis*. University of Chicago Press, 2012.

Heisenberg, Werner. *Physics and Philosophy*. Harper & Row, 1962.

Heraclitus. *The Art and Thought of Heraclitus: An Edition of the Fragments with Translation and Commentary*. Edited by Charles. H. Kahn. Cambridge University Press, 1981.

Hillman, James. *A Blue Fire.* Harper Perennial, 1989.

Hillman, James. *The Dream and the Underworld.* William Morrow, 1979.

Hinton, Geoffrey. "The so-called 'Godfather of the A.I.' joins The Lead to offer a dire warning about the dangers of artificial intelligence." Interview by Jake Tapper. *The Lead with Jake Tapper*, CNN, May 2,2023. Video, 4:27.
https://www.cnn.com/videos/tv/2023/05/02/the-lead-geoffrey-hinton.cnn

Hull, R.F.C. and McGuire, William. *C.G. Jung Speaking: Interviews and Encounters.* Princeton University Press, 1977.

Jaffé, Aniela. *The Myth of Meaning in the Work of C.G. Jung.* Daimon Verlag, 1984.

Jung, C.G. *Aion: Researches into the Phenomenology of the Self.* Princeton University Press, 1959.

Jung, C.G. *Dream Analysis 1: Notes on the Seminar Given in 1928-1930.* Princeton University Press, 1984.

Jung, C.G. *Letters: Vol. 2.* Princeton University Press, 1976.

Jung, C.G. *Psychology and Religion: East and West.* Princeton University Press, 1969.

Jung, C.G. *Psychological Types.* Princeton University Press, 1971.

Jung, C.G. *Psychology of the Unconscious.* Dover Publications, 2002.

Jung, C.G. *Synchronicity: An Acausal Connecting Principle.* Princeton University Press, 1973.

Jung, C.G. *The Collected Works of C.G. Jung, Vol. 17.* Princeton University Press, 1981.

Jung, C.G. *The Structure and Dynamics of The Psyche*. Princeton University Press, 1970.

Jung, C.G. *Two Essays on Analytical Psychology*. Princeton University Press, 1972.

Kahn, Charles H. *The Art and Thought of Heraclitus*. Cambridge University Press, 1979.

Kastner, Ruth. *Adventures in Quantumland: Exploring Our Unseen Reality*. WSPC Europe, 2019.

Kastner, Ruth. Interview by David Garofalo. "The Quantum Handshake Welcoming You to Spacetime." *David Garofalo's Corner: From Science to Culture, Black Holes to the Multiverse*. December, 24, 2020. https://www.davidgarofaloscorner.com/post/the-quantum-handshake-welcoming-you-to-spacetime-with-physicist-and-philosopher-dr-ruth-kastner

Kastner, Ruth. *The Transactional Interpretation of Quantum Mechanics: The Reality of Possibility*. Cambridge University Press, 2013.

Kastner, Ruth. Understanding Our Unseen Reality: Solving Quantum Riddles. Imperial College Press, 2015.

Kastrup, Bernardo. *Decoding Jung's Metaphysics: The Archetypal Semantics of an Experiential Universe*. Iff Books, 2021.

Kastrup, Bernardo. *Science Ideated: The Fall of Matter and the Contours of the Next Mainstream Scientific Worldview*. Iff Books, 2021.

Kelman, Harold. "Kairos: The Auspicious Moment." *The American Journal of Psychoanalysis 29* (1969): 59-83. https://doi.org/10.1007/BF01872669

Kripal, Jeffrey J., and Strieber, Whitley. *The Super Natural: A New Vision of the Unexplained.* Penguin Random House, 2016.

Kurzweil, Ray. *The Singularity is Near: When Humans Transcend Biology*. Penguin Books, 2006.

Lapp, Ralph E. "The Einstein Letter That Started It All." *The New York Times*, August 2, 1964.
https://www.nytimes.com/1964/08/02/archives/the-einstein-letter-that-started-it-all-a-message-to-president.html

Leeming, David and Margaret Leeming. *A Dictionary of Creation Myths*. Oxford University Press, 1994.

Maclagan, David. *Creation Myths: Man's Introduction to the World.* Thames and Hudson, 1979.

Martel, J.F. "Pluto and the Death of God." In *Pluto: New Horizons for a Lost Horizon*, ed. Richard Grossinger. North Atlantic Books, 2015.

Meier, C.A. Ed. *Atom and Archetype: The Pauli/Jung Letters, 1932-1958*, Princeton University Press, 2014.

Metz, Cade. "The ChatGPT King Isn't Worried, but He Knows You Might Be." *The New York Times*, March 31, 2023.
https://www.nytimes.com/2023/03/31/technology/sam-altman-open-ai-chatgpt.html

Minkowski, Hermann. "Space and Time." Lecture delivered before the Naturforscher Versammlung (Congress of Natural Philosophers) at Cologne. September 21, 1908.

Murray, Henry A. ed. *Myth and Myth Making*. Braziller, 1960.

Newton, Isaac. *Philosophiae Naturalis Principia Mathematica*. 1687

Philbrick, Ian Prasad and Wright-Piersanti, Tom. "A.I. or Nuclear Weapons: Can You Tell These Quotes Apart?" *The New York Times*, June 10, 2023.
https://www.nytimes.com/2023/06/10/upshot/artificial-intelligence-nuclear-weapons-quiz.html

Planck, Max. From a lecture entitled "Das Wesen der Materie" [The Essence/Nature/Character of Matter]. Given in Florence, Italy, 1944.

Rossi, Saffron. *The Kore Goddess: A Mythology & Psychology*. Winter Press, 2021.

Rothman, Joshua. "What Are the Odds We Are Living in a Computer Simulation?" *The New Yorker*, June 9, 2016.
https://www.newyorker.com/books/joshua-rothman/what-are-the-odds-we-are-living-in-a-computer-simulation#:~:text=Citing%20the%20speed%20with%20which,just%20%E2%80%9Cone%20in%20billions.%E2%80%9D

Rovelli, Carlo. *The Order of Time*. Riverhead Books, 2017.

Schooler, Jonathan. Interview by LD Deutsch. March 9, 2002.

Scott, Paxton, and Yin, William. "Bill Gates, Gov. Gavin Newsom speak at unveiling of new human-centered artificial intelligence institute." *The Stanford Daily*, March 19, 2019.
https://stanforddaily.com/2019/03/19/bill-gates-gov-gavin-newsom-speak-at-unveiling-of-new-human-centered-artificial-intelligence-institute

Schrödinger, Erwin. *My View of the World*. Cambridge University Press, 1964.

Schrödinger, Erwin. *What is Life & Mind and Matter*. Cambridge University Press, 1967.

Streitfeld, David. "Silicon Valley Confronts the Idea That the 'Singularity' Is Here." *The New York Times*, June 11, 2023. https://www.nytimes.com/2023/06/11/technology/silicon-valley-confronts-the-idea-that-the-singularity-is-here.html

Swetz, Frank. *Legacy of the Luoshu: The 4,000 Year Search for the Meaning of the Magic Square of the Order Three*. A.K. Peters/CRC Press, 2008.

Talbot, Michael. *Mysticism and the New Physics*. Penguin Arkana, 1993.

Toms, Michael. *An Open Life: Joseph Campbell in Conversation with Michael Toms*. Larson Publications, 1989.

Tyson, George. *Turing's Cathedral: Origins of the Digital Universe*. Vintage Books, 2012

Tyson, Neil deGrasse. *The Pluto Files: The Rise and Fall of America's Favorite Planet*. W.W. Norton, 2014.

Vallée, Jacques. "A Theory of Everything (else)" TEDxBrussels YouTube Channel, published November 23, 2011. https://www.youtube.com/watch?v=S9pR0gfil_0&lc=UgicUo6CCswWtXgCoAEC

Virk, Rizwan. *The Simulation Hypothesis*. Bayview Books, 2019.

von Franz, Marie-Louise. *Creation Myths, Revised Edition*. Shambhala Publications, 1995.

von Franz, Marie-Louise. *Number and Time*. Northwestern University Press, 1974.

von Franz, Marie-Louise. *On Divination and Synchronicity: The Psychology of Meaningful Chance*. Inner City Books, 1980.

von Franz, Marie-Louise. *Psyche & Matter*. Shambhala Publications, 2001.

von Franz, Marie-Louise. *Time: Rhythm and Repose.* Thames and Hudson, 1978.

Wheeler, John Archibald. "A Call on Physicist John A. Wheeler." *University: A Princeton Quarterly*. Summer, 1972.

Wheeler, John Archibald. "American Physicist." *Scientific American*, Vol. 267. 1992.

Yudkowsky, Eliezer. "Pausing AI Developments Isn't Enough. We Need to Shut It All Down." *Time*, March 29, 2023.
https://time.com/6266923/ai-eliezer-yudkowsky-open-letter-not-enough

Whitney, Mark. *Matter of Heart*. Kino International, 1986. 1hr, 53min.
https://www.youtube.com/watch?v=Ed3vPb9bmcw